Monographs on Astronomical Subjects: 7

General Editor, A. J. Meadows, D. Phil.,
Professor of Astronomy, University of Leicester

The Isotropic Universe

The Isotropic Universe

An Introduction to Cosmology

D. J. Raine

University of Leicester

Monographs on Astronomical Subjects: 7
Adam Hilger Ltd, Bristol

British Library Cataloguing in Publication Data

Raine, D. J.
The Isotropic Universe. — (Monographs
on Astronomical subjects, ISSN 0141-1128; 7)
1. Cosmology
I. Title   II. Series
113   QB981

ISBN 0-85274-370-X

Published by Adam Hilger Ltd, Techno House, Redcliff Way, Bristol BS1 6NX.
The Adam Hilger book-publishing imprint is now owned by The Institute of Physics.

Typeset by Unicus Graphics Ltd, Horsham, and printed in Great Britain by The Pitman Press, Bath.

To Maureen

who believes in the conspiracy theory of science

*And some poor festering Earth
Questions in a geodesic hand
The way to heaven, or the edge
Of the Universe.*

Preface

This book is intended as an introduction to cosmology at an intermediate level. It is an introduction because it assumes no knowledge of cosmology; it is intermediate because it assumes some knowledge of a lot of other things, principally first-year and some second-year university physics and a corresponding level of mathematics. But it is not concerned with the rigorous presentation of some of the more difficult arguments and results. Much of the book, with the principal exclusion of the last four chapters, is based on lectures on cosmology I have given to final-year students at Leicester. I hope it will serve as a useful bridge between the many excellent elementary texts now available and the research monographs and journals.

I have assumed also a certain basic acquaintance with astrophysics. I have a certain sympathy with those who wish to begin their study of the universe with the Universe, filling in on such details as the definition of an astronomical magnitude, or the equation of radiative transfer, or the meaning of Compton scattering as the need might (hopefully not) arise. This sympathy is deeply felt as for my own mis-spent youth. Nevertheless, mature reflection, so amply nurtured in the remoter pastures of academe, leads me to believe that this is a mistake. For cosmology is now, and has been for the last fifteen years or so, a serious branch of physics; it can, I venture to hope, be treated as such at an undergraduate level. Thus I make no further apology for the fact that the reader of this book is taken to have survived an initial acquaintance with standard topics in theoretical astrophysics, such as elementary radiative transfer, free–free emission, ionisation equilibrium, to name but a few at random, and to be familiar with basic results from thermodynamics, special relativity, electrodynamics, and elements of statistical physics and quantum theory. Indeed, it is possible to turn the argument round. Thus, the fact that cosmology draws on a wide range of physics makes it a good way of learning how many branches of the subject come together in applications once one has reached a certain basic level of competence, and this provides at least a partial justification for its inclusion in the undergraduate curriculum.

With regard to the other branch that makes a central appearance, namely general relativity, I have tried to provide an introduction sufficient for a complete understanding of the simplest cosmological models, but stopping short of the execution of more general calculations. My attitude here has been that general relativity has now become a part of physics in the same sense as, for example, electrodynamics, and there is no longer any need to exacerbate its undoubted non-triviality by an overdose of misplaced awe — misplaced because awe should be reserved for the comprehended beauty of the theory, not its mythical impenetrability. It seems that much of the mystery of general relativity could be dispelled if only the Newtonian theory of gravity were taught correctly, in terms of the Equivalence Principle, from the beginning. For, with hindsight, general relativity is a natural generalisation of the suitably formulated Newtonian theory of gravity to the relativistic context.

Despite the apparent specialisation indicated in the title of this book, I am concerned, to borrow a phraseology, not with an aspect of cosmology, but with the whole of cosmology in one aspect. Until the last fifteen years, the canonical title for such a book would have been something like 'The Expanding Universe', emphasising the central role played by the discovery of expansion. The Hubble law still remains the single most important discovery in cosmology, but I think that it is now appropriate to emphasise the remarkable degree of isotropy, with the implied large-scale uniformity, as the central observational feature. The general theory of relativity then provides models in which the distant galaxies must exhibit redshifts, as observed, and these can be interpreted in a well defined manner in terms of a universal expansion. What I am trying to emphasise is that the naive Doppler interpretation of the redshifts has no place in the scheme of things. The *isotropic* universe therefore is an emphasis not an exclusion. The dramatic nub of this viewpoint is the way in which the high degree of isotropy develops as a paradox as a consequence of itself.

Turning now to some technicalities, I have used the MKS system of units as standard. Readers who stray into the cosmological literature will rapidly realise that nowhere are MKS units to be found, and that they had better familiarise themselves with the CGS system. I have not agonised much over this, since the conversion factors, outside of electrodynamics, are trivial. There is one exception to this: it is usual in relativity to use units in which the speed of light, c, is put equal to unity so that distances are expressed in light seconds. Many beginning students find this confusing, and for the most part I have left c in equations. However, in order to avoid factors of c in metric expressions I found it efficient to employ rather unconventional units in which time is measured in 'light metres'. To get back to standard units one makes the replacement $t \rightarrow t/c$.

My policy on giving references has been to keep them to a minimum, since this leads to a list which is quite long enough. Wherever appropriate, I have given references to secondary sources; I have omitted references which can be readily found in such sources, references to material that is standard, and to lines of investigation which I regard as essentially tangential. I have also ignored all of these guidelines wherever I felt it to be appropriate.

Many people have contributed in varying degrees to this book, many unwittingly, one anonymously, none culpably, all appreciated. I am grateful to those who only waited, for their patience and its opposite (which, in these circumstances, is usually referred to as encouragement), while the slow dredge of authorship inked and re-inked an impenetrable maze to be transcribed by a valiant typist, and to the publishers for the courtesy and efficiency with which they dealt with a similarly reprocessed typescript. And I should like to thank a teacher, to whom I owe my introduction to cosmology, and my students, who have tried to teach me how to teach, in the hope that their efforts have not been entirely in vain.

It is too much of a cliché to express the hope that this book will encourage a handful of its readers to further study and insight. In fact, the result I think I should really like is that the readers of this book should find it entertaining. For to be entertained is, I suppose, the only sane response to a Universe so scaffolded in misconception.

D. J. Raine
University of Leicester

Contents

1.   *The Quality of Matter*

1.1. The Patterns of the Stars

It is difficult to resist the temptation to organise the brightest stars into patterns in the night sky. Of course, the traditional patterns of various cultures are entirely different, and few of them have any physical significance; most of the patterns visible to the naked eye are mere accidents of superposition; their description and mythology represent nothing more than Man's desire to organise his observations while they are yet incomplete . . .

In the scientific study of cosmology we are not interested primarily in individual objects but in the statistics of classes of objects. From this point of view, more important than the few brightest stars visible to the naked eye are the statistics written in the band of stars of the Milky Way. Studies of the distribution and motion of these stars reveal that we are situated towards the edge of a rotating disc of some 10^{11} stars, 30 kpc in diameter, which we call the Galaxy (figure 1.1). Physically, the stellar types are characterised by mass, initial chemical composition

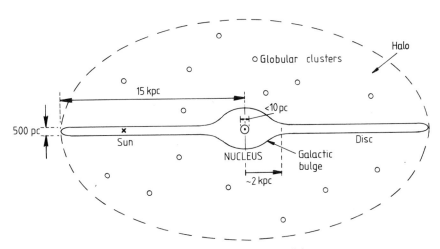

Figure 1.1. Schematic view of the Galaxy.

1

and age; observationally, they are classified by spectral type, or surface temperature, and luminosity class ranging from supergiant (I) to dwarf (V). The main chemical and kinematical subclasses are described in terms of stellar populations. The young ($<5 \times 10^8$ years), metal-rich, low-velocity stars of Population I trace out a spiral structure in the disc. This spiral structure is also apparent in the 21 cm radio emission from neutral hydrogen. The bulk of the mass of the Galaxy (about 70%) belongs to the old disc population with solar metal abundances, intermediate peculiar velocities and intermediate ages. The oldest high-velocity stars of Population II form a spheroidal distribution about the Galaxy; their metal abundances are as low as 1% of the solar value. Our view to the Galactic centre is obscured at optical wavelengths because the starlight is scattered by micromillimetre-sized grains of dust, of uncertain composition, associated mainly with the spiral arms. This extinction is important for extragalactic observations since it varies with Galactic latitude. Extragalactic observations have to be made away from the plane of the Galaxy and corrected for the reddening of the light associated with this wavelength-dependent scattering. Even then interpretation of observations can be complicated by the patchiness of the absorption. At infrared and radio wavelengths the scattering is much less and emission from the Galactic centre can be detected. On the short-wavelength side, obscuration due to photoelectric absorption by neutral hydrogen falls off again above medium-energy x-rays ($\gtrsim 2$ keV).

Distributed around the Galaxy in an approximately spherically symmetrical manner, and moving in elliptical orbits about the Galactic nucleus, are some 200 globular clusters. These are dense spherical associations of 10^5–10^7 Population II type stars within a radius of 10–20 pc.

The Andromeda Nebula (M31), which, at a distance of 660 kpc is the furthest object visible to the naked eye, provides us with a view of how our own Galaxy must look to an Andromedan astronomer. The collection of our near-neighbour galaxies is called the Local Group. Andromeda and the Galaxy are dominant among the 20 or so members of the Group. These include the dwarf elliptical galaxies Fornax and Sculptor, the dwarf irregular Larger and Smaller Magellanic Clouds, the spiral galaxy M33, and the elliptical M32.

If we were able to turn up the contrast of the night sky so that sources of apparent magnitude brighter than $m_v = +13$ became visible, then, ignoring foreground stars, we should be able to pick out a band of galaxies across the sky in the direction of the giant elliptical galaxy M87 in Virgo. This galaxy lies at a distance of about 11 Mpc and is at the centre of a rich, irregular cluster of some 2000 member galaxies called the Virgo cluster. Many, if not all, of the galaxies in this band are members of groups of galaxies of various sizes, of which our Local Group is a relatively small example. Centred on the Virgo cluster, the distribution is referred to as the Virgo supercluster or Local Supercluster. Again we appear to live near the edge of this system. It appears that the Local Supercluster is flattened rather than spherical; the extent to which such flattening can be due to a possible rotation of the system is a matter of debate.

There are many other examples of rich clusters of galaxies, of which the best known and most studied is the Coma cluster. These have been catalogued and investigated to varying degrees, as have the many less spectacular groupings. There is evidence of a tendency towards clustering of the rich clusters themselves with perhaps 10% of them belonging to superclusters containing ten or so rich clusters as members (in addition, of course, to many more galaxies not in rich clusters). We shall return to this in more detail later (see §1.5), but for the moment the point is that there is definite evidence for a pattern in the distribution of matter on scales of some 20 Mpc. In order to discover how this pattern continues to larger scales, we have to analyse in some detail the distribution of galaxies. In order to understand the significance of what is revealed by this analysis, we first look at the distributions that might be found.

The average density of matter in the Galaxy is about 2×10^{-21} kg m^{-3}, obtained by dividing the total mass by the volume of the disc. If we average instead over the volume of the Local Group, which we might regard as a typical region of the Local Supercluster, we obtain a density of 0.5×10^{-25} kg m^{-3}. The average density of a rich cluster is approximately 2×10^{-24} kg m^{-3}, whereas that of a typical supercluster may be 2×10^{-26} kg m^{-3}. The average density here clearly depends on both the size and location of the volume which is being averaged over. If we ignore the atypical result for rich clusters, which contain only about 10% of galaxies, then the trend is towards a decrease in mass density the larger the sample volume. The same trend can be observed for the number density of galaxies, which for this purpose can be assumed not to vary in mass. It is possible that the mass density, or the number density, does not have a limit but, for example, oscillates with finite amplitude as we go to scales larger than superclusters. This would represent a rather strange distribution of matter and we shall not discuss it further. The two remaining distinct possibilities are that the density tends to a limit which is either finite or zero.

We can arrange for the limiting density to be zero in an infinite universe if we assume an infinite hierarchy of clustering: clusters of order n are themselves clustered to form a cluster of order $n + 1$ ($n = 1, 2, \ldots$). The clusters of order $n + 1$ within a cluster of order $n + 2$, are taken to be separated by a distance much larger than n times the separation of clusters of order n within an $(n + 1)$th-order cluster. Such a distribution is called a hierarchical model of the Universe. In such a system the concept of an average density is either meaningless, or useless, since the density depends on the volume of space averaged over, except in the limit, when its zero value tells us nothing.

Suppose, on the other hand, that the Universe consists of randomly arranged clusters of some particular order m which are themselves therefore not clustered. Of course, the random arrangement will produce some accidental groupings of mth-order clusters; by saying that there are no $(m + 1)$th-order clusters we mean that these groups of mth order clusters occur no more frequently than expected for a *random* distribution. Provided we average over a volume containing many

3

*m*th-order clusters, then, apart from statistical fluctuations, we shall obtain a result for the mean density of matter which is independent of location and volume. Looked at on a large enough scale such a system appears uniform, just as the randomly aggregated grains of an amorphous solid give rise to a macroscopically uniform material. A universe with this property is called *homogeneous*. Note that we are here neglecting the expansion of the Universe (Chapter 2) which would mean that such a universe would not appear uniform to a given observer looking back in time (see §§1.6, 6.3 and 8.9). An alternative way of arriving at a homogeneous universe is to arrange that some, but not all, clusters of order $m - 1$ are grouped to give clusters of order m, and that the remainder are instead randomly distributed. The application of our mathematical analysis in §1.10 will show that the Universe is apparently homogeneous on a sufficiently large scale, but that neither of these suggestions as to how this is achieved quite matches the reality.

1.2. The Edge of the Universe

The simplest question one can pose is to ask whether the distribution of galaxies has an edge analogous to the boundary of the distribution of stars in the Milky Way. The first attempt to answer this question was made by Hubble around 1934. He was able to probe the large-scale distribution of galaxies beyond the Local Supercluster by counting galaxies down to an apparent photographic magnitude $m \approx 21$. This corresponds to a visual magnitude $m_v \approx 19.8$ it we neglect certain possible small corrections. From this, and the knowledge of the typical absolute magnitude of galaxies, we can estimate the distance to which Hubble was able to survey. By considering the number of galaxies $N(<m)$ brighter than apparent magnitude m, we shall then show that, to the depth of his survey, Hubble found no evidence of an edge to the galaxy distribution.

First we need the typical absolute magnitude of galaxies. The integrated galaxy luminosity function $\Phi(M)$ is shown in figure 1.2. This gives the total number of galaxies per unit volume with absolute (visual) magnitude greater than M. Thus $d\Phi/dM . dM$ is the number of galaxies per unit volume with magnitudes between M and $M + dM$. The luminosity function is derived from studies of nearby galaxies, the distances to which have been determined by one of the methods of Chapter 2. In effect, therefore, we use the distance information on nearby galaxies to determine their distribution in luminosity, and then use the apparent luminosity of more distant galaxies to determine their distribution in distance. This assumes that there is no systematic evolution of the luminosity function, an assumption which is completely reasonable as a first approximation for the limited depth of the galaxy surveys.

There is a characteristic break in the luminosity function at visual magnitude $M^* = -19.5$. This value is arrived at for a distance to the Coma cluster of 70 Mpc, which is the calibration of the distance scale we shall use throughout this book; if the distance to Coma were in fact $70/h$ Mpc we should have $M^* = -19.5 + 5 \log h$.

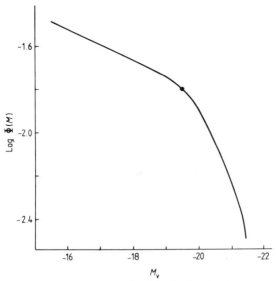

Figure 1.2. The form of the integrated galaxy luminosity function $\Phi(M)$, giving the number of galaxies per Mpc³ having absolute magnitudes brighter than M.

For a rough estimate of the depth of Hubble's survey we may assume that the total contribution to the counts of galaxies comes from those with $M_V = M^*$. This means that we neglect brighter galaxies because the rapid decline in $\Phi(M)$ means there are relatively less of them, and we neglect fainter galaxies because they are less likely to be detected. With this assumption, and the definition of absolute magnitude

$$m_V - M_V = 5 \log d - 5,$$

where d is in parsecs (pc), we deduce that $m_V = 19.8$ corresponds to a distance of about $550/h$ Mpc. This is clearly only a rough approximation; Peebles (1971) gives a more accurate calculation which yields a result a factor 2 larger.

The idea now is to estimate the function $N(<m)$ by counting numbers of galaxies as a function of apparent magnitude. Assume first, for simplicity, that all galaxies have the same absolute magnitude M^*. Suppose that in the region surveyed the distribution is uniform with galaxy number density n per unit volume. Then the number of galaxies counted brighter than magnitude m equals the number within a sphere of radius r_{M^*} pc determined by the fact that a galaxy at the limiting distance r_{M^*} has apparent magnitude m, i.e.

$$m - M^* = 5 \log r_{M^*} - 5. \tag{1.2.1}$$

From this, and

$$N(<m) = N(<r_{M^*}) = \tfrac{4}{3} \pi r_{M^*}^3 \, n,$$

we obtain

$$\log N(<m) = 0.6 \, m + \text{constant}. \tag{1.2.2}$$

5

To take account of the different intrinsic luminosities of galaxies, we integrate over all luminosities; thus

$$N(<m) = \int \frac{d\Phi}{dM} \, dM \, \tfrac{4}{3}\pi r_M^3$$

and, using (1.2.1) this again yields a relation of the form (1.2.2). Any departure from uniformity will appear in a plot of $\log N(<m)$ against m as a departure from a straight line of slope 0.6. The observed $N(<m)$ must be corrected for galactic absorption.

Hubble found the number of the faintest galaxies to be about one-half the value predicted here on the assumption of homogeneity and Euclidean geometry. This is probably due to a systematic error in estimating magnitudes of faint galaxies. The correct conclusion from the work is that, to within a variation in density of a factor 2, Hubble did not find an edge to the distribution of galaxies out to about 1000 Mpc. Thus there is no evidence for clustering with a density enhancement of greater than a factor 2 on this scale. To a first approximation we may regard the distribution as homogeneous on a large scale.

A further important result from Hubble's survey concerns the dependence of the galaxy distribution on direction in the sky. Hubble actually looked at small regions of the sky in different directions; no significant difference was found between these regions. A universe which looks the same in all directions from a given point is said to be *isotropic* about that point. Thus Hubble's results show that the galaxies are distributed at least approximately isotropically about us. We may conclude that a model of the Universe with a uniform distribution of matter is a suitable first approximation.

1.3. Homogeneity and Isotropy

A uniform distribution in space is said to be homogeneous, and a distribution uniform with respect to direction about us is said to be isotropic. The relation between these notions and their meaning can be highlighted by considering what the opposite type of inhomogeneous or anisotropic universes might look like.

It is possible for a system to be homogeneous but not isotropic about any point, since at each point there may be a privileged direction. A uniform magnetic field provides a simple example. In terms of distributions of galaxies one can imagine an anisotropic universe in which the average separation of galaxies is smaller in one direction than in others, but which is uniform in the sense that a neighbourhood of one point looks the same as a corresponding neighbourhood of any other point. An inhomogeneous system may be isotropic about one or more points. For example, a solid sphere is isotropic about its centre if its density is a function of radius only. From this it follows that the isotropy of the galaxy distribution about us does not imply isotropy about any other point, unless we assume that we do not occupy a privileged location in the Universe. On the other hand, a system which is isotropic about every point is necessarily homogeneous. For if the density of galaxies $n(x)$ were inhomogeneous this would single out at each point a direction in which the gradient of $n(x)$ would be steepest, thereby implying anisotropy. Therefore, the observation of isotropy of the Universe about us, together with the assumption that we are not special observers, enables us to infer isotropy about every point, and hence homogeneity.

An example of an inhomogeneous universe is provided by a cloud of galaxies of finite extent with empty space beyond. If the galaxies were distributed uniformly within the cloud then for an observer situated well away from the edge the observed distribution would be indistinguishable from an infinite homogeneous universe. Fortunately we can rule out such a model on other grounds (§3.3). The hierarchical universe already considered (§1.1) is an example of an infinite inhomogeneous universe.

There is substantial evidence, some of which we have already considered and the remainder of which we shall discuss in due course, that the Universe is homogeneous and isotropic on a large scale. On smaller scales, up to that of the Local Supercluster and possibly larger, we know that inhomogeneities occur. A discussion of the distribution of matter requires a discussion of the nature of these inhomogeneities. There are two complementary ways in which this is to be done. On the one hand, we try to categorise the types of inhomogeneity by giving names to those we recognise as qualitatively different, in a sort of celestial botany: groups, rich clusters, clouds of clusters, superclusters, etc, are examples of such a classification. Such objects are then subject to detailed study and statistical analysis. On the other hand, we can try to express the statistical nature of the distribution of galaxies by providing statistical measures of the type of distribution with which we are dealing: clusters might arise from purely statistical fluctuations in a random distribution of galaxies (although in fact they do not); or we might try to measure the departure of the distribution from a purely random one. Of course, we could do this precisely by giving the exact location of every galaxy in the visible Universe. In principle, this contains all the relevant information, and there is no need to say more about either the classification or statistical description of inhomogeneity. In practice, however, a list of 3×10^9 coordinates tells us nothing of assimilable interest, so we look for simple statistics to characterise the data.

It is this second approach that we describe next (§§1.5-1.12), although the first will be taken up again briefly in §1.13. It is important to note that it is assumed in the analysis that the Universe is homogeneous and isotropic in the large, and that our task is to describe localised departures from homogeneity. In as far as the analysis yields internally consistent and meaningful results, this provides a check on the assumptions of homogeneity and isotropy.

1.4. Catalogues of Galaxies

An early galaxy catalogue, still of some importance, is the Shapley-Ames (1932) catalogue giving the coordinates and magnitudes of 1250 galaxies brighter than

13th magnitude over the whole sky. This has been extended in the *Reference Catalogue of Bright Galaxies* (de Vaucouleurs and de Vaucouleurs 1964). From the Palomar Observatory *Sky Survey* photographs, Zwicky *et al* (1961-68) have prepared a catalogue of about 5000 galaxies brighter than 15th magnitude in the northern hemisphere. This corresponds to a distance of the order of 100 Mpc. The catalogue of Shane and Wirtanen (1967), based on the *Lick Astrographic Survey*, contains some 10^6 galaxies to a limiting magnitude of 19 or a distance of approximately 500 Mpc. For such a large number of galaxies it is impractical to give individual coordinates. Instead, the sky is divided into $10' \times 10'$ cells and the number of galaxies in each cell is recorded. An average cell contains one galaxy, but there are substantial deviations from average; for example, there are more than ten galaxies in 6600 of the cells. An even deeper survey has been carried out by Rudnicki *et al* (1973) of the Jagiellonian University in Cracow. This covers a $6° \times 6°$ area of the sky and extends to magnitude 20.5; some 10 000 galaxies are included.

A catalogue of clusters of galaxies has been prepared by Abell (1958) from the Palomar *Sky Survey* plates. This lists 2712 of the richest clusters of galaxies, of which the Coma cluster is an extreme example, to a depth of about 600 Mpc over a large fraction of the sky. These clusters are often referred to by their Abell catalogue numbers; for example, Coma itself is A 1656. The Zwicky (1961-68) *Catalogue of Galaxies and Clusters of Galaxies*, prepared from the same plates, is a more extensive compilation. Since different criteria for cluster membership are employed, the Abell and Zwicky catalogues are not directly comparable.

All of these catalogues require the identification of galaxy images on the photographic plates. While it is easy to distinguish galaxies from foreground stars by the fuzzy nebulosity of the galaxy images, the process is nevertheless exceedingly time-consuming. This problem should be considerably alleviated with the current introduction of automated galaxy recognition and positional recording devices (e.g. Pratt 1977).

1.5. Clustering of Rich Clusters

Consider first the distribution of rich clusters of galaxies. Even if the clusters are distributed at random a certain amount of accidental clustering will occur. We therefore want to know if the clusters exhibit any additional clustering tendency, so we must compare the observed distribution with a random scattering of cluster locations. To do this we divide the sky into cells and count the number of clusters in each cell.

For clusters distributed at random with an *average* number of clusters per cells n, the probability of finding m clusters in a cell approximates a Poisson distribution, given by

$$P(m) = n^m \, e^{-n}/m!$$

If the distribution is really random we therefore expect to find cells with m clusters occurring $NP(m)$ times, where $N(\gg m, n)$ is the total number of cells. We can compare this expected number, $E_m = NP(m)$, on the hypothesis of a random distribution, with the observed numbers O_m from the galaxy catalogue. The sum

$$\chi^2 = \sum_{m=1}^{N} \frac{(O_m - E_m)^2}{E_m}$$

is a measure of the departure of the actual distribution from the expected one. Of course some difference from the average behaviour is to be expected in the single sample provided by a small part of our one Universe, so a non-zero value of χ^2 is not necessarily incompatible with the hypothesis of a random distribution of clusters. Tables of χ^2_{N-1} give the probability $P(\chi^2)$ of any particular value of χ^2 arising from a purely statistical fluctuation in a sample of N clusters from a random distribution. For the distribution of Abell clusters $P(\chi^2)$ turns out to be between 10^{-30} and 10^{-40}, depending on the number of cells chosen (Abell 1975), which means that the probability of the catalogued distribution arising from a random distribution of clusters is 1 in 10^{30}–10^{40}. This provides apparently strong evidence that the distribution of rich clusters on the sky is not random!

In fact the result is not quite as surprising as it at first seems. We know that galactic obscuration must vary across the sky and this could account for a large part of the departure from randomness. Clearly a more detailed analysis is required.

For this purpose the clusters are divided into six classes according to their apparent luminosity, which, it turns out, is equivalent to a classification according to their distance. The clustering within each distance class can then be analysed. What is required is an analysis based on relative values of $P(\chi^2)$ rather than the absolute value, and this is provided by investigating the way $P(\chi^2)$ varies with cell size within each distance class. For very small cells, almost all cells contain one cluster or none, and this provides no information; for very large cells the departure from randomness must reflect just galactic obscuration. Thus the cell size which gives the maximum departure from randomness, or minimum $P(\chi^2)$, is of special interest, since this is presumably the scale of superclustering of clusters.

If the superclustering is a real effect then one expects it to be independent of distance class. The more distant clusters should therefore give an angular scale of superclustering which is smaller than that for nearby clusters in inverse ratio to their distances. This effect is indeed observed. Furthermore, galactic (or intergalactic) obscuration cannot account for all of the effect since clusters of distance class 6, the most distant ones, are observed in regions deficient in nearer clusters. The results indicate that perhaps about 10% of very rich clusters belong to superclusters with an average of ten members (which, presumably, contain in addition many less rich clusters).

Statistical tests based on the division of the sky into cells may be subject to systematic errors due, for example, to large-scale variability in Galactic absorption, or the variation of zenith distance at which plates are taken. The χ^2 test shows that

9

the distribution of galaxies is not random on *any* scale, presumably as a result of systematic errors. It is therefore not clear that the scale on which departures from randomness is a maximum is really to be identified as the scale of superclustering, rather than with some aspect of the systematic errors. An alternative mode of analysis, which overcomes this difficulty, is provided by the method of power spectra (Blackman and Tukey 1959). In the next sections we describe this method to the point where we can show how it is used to confirm the superclustering of rich clusters, and how it is applied to analyse the clustering of galaxies themselves.

1.6. The Correlation Function and Power Spectrum Analysis

We assume that the objects we are analysing (galaxies or clusters of galaxies) can be regarded as point particles, and that these are distributed homogeneously on a sufficiently large scale; in particular, we assume that we can meaningfully assign an average number density. If we make this assumption, we can characterise the galaxy distribution in terms of the extent of the departures from uniformity on various scales. The following analysis is inapplicable if this assumption is not justified. For simplicity we shall also assume here that we are dealing with a static universe and so neglect the effects of expansion (Chapter 2). Since the distance surveyed in the deepest catalogue is about 1000 Mpc, the light travel time from the most distant galaxies of the samples is less than 30% of the age of the Universe, so the neglect of expansion is appropriate in a first approximation. Recent work which has included the effects of expansion has shown improved agreement between theory and observation in the scaling laws (§1.8). We consider first the three-dimensional distribution of galaxies. The catalogues of galaxies contain no distance information, and so yield directly only a distribution on the celestial sphere. The relation between this two-dimensional projection and the distribution in depth will be discussed in §1.8.

If the average number density of galaxies is $\bar{n}$ then we have to go on average a distance $\bar{n}^{-1/3}$ from a given galaxy before we encounter another. We can describe the local departures from uniformity by specifying the distance one actually has to go from any particular galaxy. Sometimes this will be larger than average, sometimes less. To specify the distance in each case is equivalent to giving the locations of all galaxies, which is not the sort of description we are after. What we require is a statistical description giving the number of times one need only travel a certain fraction of the average distance to find a nearest neighbour, or, equivalently, the probability of finding a nearest neighbour within a certain distance. This information is contained in the (two-point) correlation function $\xi(r_1, r_2)$, to be defined below.

Now the probability of finding a galaxy closer than 50 kpc to the Milky Way is zero, and within a distance greater than 60 kpc it is unity (since this includes the Magellanic Clouds). This is clearly not the sort of probability information we have in mind; what we need is some sort of average. The actual Universe may be thought of as one particular realisation of some statistical distribution of galaxies. We may therefore imagine many universes each constructed according to the same statistical law, but each with a different realisation in detail of that law. What these universes have in common expresses the content of the law, and what they have in common is expressed through quantities averaged over the 'ensemble' of universes. In each realisation, therefore, the points to be thought of as representing the galaxy and the LMC will occur at different distances. The departure from randomness due to clustering will be represented by the fact that the *average value* over the ensemble of this separation is less than $\bar{n}^{-1/3}$.

In practice, we do not have at our disposal an ensemble of different universes, and we can only construct such an ensemble theoretically if we already know the statistical law we are in fact trying to find. In practice, therefore, we take a spatial average over the visible Universe, or as much of it as has been catalogued, in place of an ensemble average. This makes sense if the departure from uniformity occurs on a scale less than the depth of the sample, so that the sample adequately reflects the *statistical* properties of the Universe as a whole. The results suggest that this condition is reasonably fulfilled in practice.

For a completely random but homogeneous distribution of galaxies, the probability dP_1 of finding a galaxy in an infinitesimal volume dV_1 is proportional to dV_1 and to the average number density of galaxies, $\bar{n}$, and is independent of position:

$$dP_1 = (\bar{n}\, dV_1)/N,$$

where N is the total number of galaxies in the sample. As we have explained, the meaning of the probability here is an average over the galaxy sample: we divide space into volumes dV_1 and count the ratio of those cells which contain a galaxy to the total number. The probability of finding two galaxies in a cell is of order $(dV_1)^2$, and so can be ignored in the limit $dV_1 \to 0$. This procedure makes sense if the galaxies are distributed randomly and uniformly on some scale less than that of the sample.

If galaxies were not clustered, the probability dP_{12} of finding galaxies in volumes dV_1 and dV_2 would be just the product $dP_1\, dP_2$ of the probabilities of finding each of the galaxies, since in a random distribution the positions of galaxies are uncorrelated. Any departure from a random distribution means that the joint probability differs from a simple product, and this difference defines the two-point correlation function $\xi(\mathbf{r}_1, \mathbf{r}_2)$. By definition, we put

$$dP_{12} = (\bar{n}^2/N^2)\,[1 + \xi(\mathbf{r}_1, \mathbf{r}_2)]\, dV_1\, dV_2 \qquad (1.6.1)$$

for the probability of finding a pair of galaxies in volumes dV_1, dV_2 at positions $\mathbf{r}_1, \mathbf{r}_2$. Of course, our assumption of randomness on sufficiently large scales means that $\xi(\mathbf{r}_1, \mathbf{r}_2)$ must tend to zero if $|\mathbf{r}_1 - \mathbf{r}_2|$ is sufficiently large. Also, the assumption of homogeneity means that ξ cannot depend on the location of the galaxy pair but only on the distance $|\mathbf{r}_1 - \mathbf{r}_2|$ separating them; for the probability must be independent of the location of the first galaxy. If ξ is positive we have an excess probability over a random distribution and hence clustering; if ξ is negative we have anti-clustering. Clearly ξ must lie in the range $-1 < \xi < \infty$. There is an obvious extension of the definition to the n-point correlation functions which will be a function of $n-1$ relative distances. In practice, computations have not been carried out beyond the four-point function.

For the purpose of calculation it is often useful to replace the description in terms of point particles by a continuum description. We picture the galaxies as the microscopic constituents of a continuous fluid with variable density $n(\mathbf{R})$, at $\mathbf{R}$. Thus we imagine an averaging to have been carried out over scales large compared to galactic separations, but small compared to clustering scales. Averaging over a volume, V, which is large compared to the scale of clustering, this gives

$$\int_V n(\mathbf{R})\, dV = \bar{n}V, \tag{1.6.2}$$

where dV is an element of volume at $\mathbf{R}$. The joint probability of finding a galaxy in dV_1 at $\mathbf{R} + \mathbf{r}_1$ and in dV_2 at $\mathbf{R} + \mathbf{r}_2$ is given by

$$n(\mathbf{R}+\mathbf{r}_1)\, n(\mathbf{R}+\mathbf{r}_2)\, dV_1\, dV_2/N^2.$$

Averaging this over the sample gives

$$dP_{12} = \frac{1}{N^2 V} \int_V n(\mathbf{R}+\mathbf{r}_1)\, n(\mathbf{R}+\mathbf{r}_2)\, dV\, dV_1\, dV_2.$$

Comparing this with (1.6.1), and writing τ for $\mathbf{r}_2 - \mathbf{r}_1$, we obtain

$$\bar{n}^2[1+\xi(\tau)] = 1/V \int_V n(\mathbf{r})\, n(\mathbf{r}+\tau)\, dV, \tag{1.6.3}$$

where $\mathbf{r}$ has been written for $\mathbf{R}+\mathbf{r}_1$, and dV is now an element of volume at $\mathbf{r}$. An alternative form for (1.6.3) is

$$\bar{n}^2\xi(\tau) = 1/V \int_V [n(\mathbf{r})-\bar{n}]\,[n(\mathbf{r}+\tau)-\bar{n}]\, dV, \tag{1.6.4}$$

which is readily verified on evaluating the right-hand side term by term using (1.6.2). This shows how $\xi(\tau)$ is related to the variance of the distribution; in fact, the variance is $\bar{n}^2\xi(0)$.

Related to the correlation function is the so-called power spectrum of the distribution. This is the Fourier transform of the correlation function defined by the integral

$$P(\mathbf{k}) = (2\pi)^{-3/2}\bar{n} \int \xi(\tau) \exp(i\mathbf{k}\cdot\tau)\, d^3\tau. \tag{1.6.5}$$

Since $\xi(\tau)$ is a function of $\tau = |\tau|$ only, the integral can be simplified by choosing polar coordinates with polar axis along k, and explicitly performing the integration over angles. This yields

$$P(k) = \left(\frac{2}{\pi}\right)^{1/2} \bar{n} \int_0^\infty \xi(\tau) \left(\frac{\sin k\tau}{k\tau}\right) \tau^2 \, d\tau,$$

with $k = |\mathbf{k}|$. The only property of the Fourier transform that we need, apart from the definition (1.6.5), is the inversion formula which enables $\xi(\tau)$ to be reconstructed from a knowledge of $P(k)$; for a sample of finite rectangular volume V, this is

$$\xi(\tau) = (2\pi)^{-3/2}(\bar{n}V)^{-1} \sum_{k \neq 0} P(k) \exp(-i\mathbf{k} \cdot \tau),$$

where the sum is over plane waves periodic in the volume V.

The galactic fluid can be thought of as composed of waves of density with various wavenumbers $\mathbf{k}$. The power spectrum gives the intensity or power in each component. To see this, we can perform the Fourier decomposition of the density in the volume V:

$$n(\mathbf{r}) - \bar{n} = \sum_{k \neq 0} \exp(i\mathbf{k} \cdot \tau) n_k,$$

the sum again being over waves periodic in V. The $k = 0$ term is excluded by the condition that both sides of the equation have zero mean. Then, denoting average values by $\langle \rangle$,

$$\bar{n}^2 \xi(\tau) = \langle [n(\mathbf{r}) - \bar{n}][n(\mathbf{r} + \tau) - \bar{n}] \rangle$$

$$= \left\langle \sum_{k \neq 0} \exp(i\mathbf{k} \cdot \mathbf{r}) n_k \sum_{k' \neq 0} \exp[i\mathbf{k}' \cdot (\mathbf{r} + \tau)] n_{k'} \right\rangle$$

$$= \sum_k \sum_{k'} \langle \exp[i(\mathbf{k} + \mathbf{k}') \cdot \mathbf{r}] \rangle n_k n_{k'} \exp(i\mathbf{k}' \cdot \tau).$$

The average of the exponential is zero unless $\mathbf{k} = -\mathbf{k}'$, so we get

$$\bar{n}^2 \xi(\tau) = \sum_{k \neq 0} n_k n_{-k} \exp(-i\mathbf{k} \cdot \tau).$$

Now $n_k^* = n_{-k}$ follows from the condition that $n(\mathbf{r}) - \bar{n}$ be a real number; for then

$$\sum \exp(-i\mathbf{k} \cdot \mathbf{r}) n_k^* = \sum \exp(i\mathbf{k} \cdot \mathbf{r}) n_k = \sum \exp(-i\mathbf{k} \cdot \mathbf{r}) n_{-k}.$$

We obtain

$$\bar{n}^2 \xi(\tau) = \sum_k |n_k|^2 \exp(-i\mathbf{k} \cdot \tau),$$

and hence

$$P(k) = (2\pi)^{-3/2}(\bar{n}V) |n_k|^2 / \bar{n}^2. \tag{1.6.6}$$

13

Since the power spectrum and the correlation function are simply related by a Fourier decomposition, the two functions must contain precisely the same information and hence be equivalent. This would indeed be so if the functions could be obtained exactly. However, at best we have only a finite amount of data and those data may be subject to systematic error. This means that the functions must be estimated from finite data, and contain information not only on real physical effects, but also on unknown errors. These errors may appear in the results in different ways for the two functions, and the extraction of the physically significant statistical information may be more easily accomplished from one or the other function. The usual approach is to try both and see. For example, there will be practical errors arising from variations in sensitivity of photographic plates. The small-scale behaviour of the correlation function is affected by errors on all scales, whereas the power spectrum depends only on the wavelength of its argument. Thus a systematic variation across a plate will affect $P(k)$ only at large wavelengths corresponding to the scale of the plate, so this may provide a better representation of the data for the purpose of extracting such an effect.

1.7. Examples of Correlation Functions

1.7.1. A One-dimensional Example

Suppose we have points distributed on a line in non-overlapping clumps of length a, with average density $\bar{n}$, and constant density within the clumps n_c. Furthermore, suppose the clumps are themselves distributed at random on the line. This is a model for a situation in which all galaxies occur in clusters which are randomly distributed. We work out $\xi(\tau)$ for a sample length of data L in the following way. The average value of the product $n(x)\,n(x+\tau)$, for a fixed τ, is given by summing over positions of x in cells of length δL according to the standard definition of an average:

$$\langle n(x)\,n(x+\tau)\rangle = \sum_{x \in \delta L} (\text{probability that } x \text{ is in } \delta L)$$
$$\times (\text{value of } n(x)\,n(x+\tau) \text{ if } x \text{ is in } \delta L).$$

If $\tau > a$, we obtain a non-zero contribution to the sum only if x lies in one clump and $x+\tau$ lies in another. For a large number of clumps the probability of this is approximately $(L_c/L)^2$, where L_c is the total length of the clumps, since the clumps are randomly distributed. Thus $\langle n(x)\,n(x+\tau)\rangle = (L_c/L)^2 n_c^2$. To compute L_c, note that the average density is given by $\bar{n} = n_c L_c/L$; so for $\tau > a$,

$$\langle n(x)\,n(x+\tau)\rangle = \bar{n}^2.$$

From (1.6.3) we obtain $\xi(\tau) = 0$ for $\tau > a$; this is correct since the location of a second point outside a cluster is random.

14

If $\tau < a$, we get two contributions to the sum: either (i) x and $x + \tau$ lie in the same clump; or (ii) they belong to different clumps.

(i) The total probability that x and $x + \tau$ belong to the same clump is the probability that x lies in the interval $[0, a - \tau]$, multiplied by the number of clumps; i.e. the probability that x lies in $(a - \tau) L_c/a$ of the sample length. This contribution is therefore

$$n_c^2 (a - \tau) \frac{L_c}{a} \frac{1}{L} = \bar{n} n_c (1 - \tau/a).$$

(ii) This is possible if x lies within τ of the end of a clump and $x + \tau$ lies in a clump. The probability of this is approximately

$$\left(1 - \frac{a - \tau}{a}\right) L_c \frac{1}{L} \frac{L_c}{L},$$

and hence (ii) gives a contribution

$$n_c^2 \frac{\tau}{a} \frac{\bar{n}^2}{n_c^2} = \bar{n}^2 \tau/a$$

to the sum. From equation (1.6.3) we have

$$\bar{n}^2 [1 + \xi(\tau)] = \bar{n} n_c (1 - \tau/a) + \bar{n}^2 \tau/a,$$

and hence

$$\xi(\tau) = (n_c/\bar{n} - 1)(1 - \tau/a) \qquad (\tau < a).$$

Note that $\xi(\tau)$ goes smoothly to zero as $\tau \to a$, in accordance with the existence of clustering only on scales $< a$. The precise shape of the function $\xi(\tau)$ for $\tau < a$ depends on the density within each cluster. For realistic density distributions within the clumps, and in three dimensions, the correlation function exhibits a characteristic knee (figure 1.3).

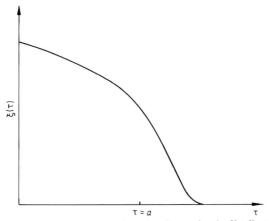

Figure 1.3. A schematic correlation function for randomly distributed clusters of scale a.

15

The characteristic length scale is again evident if we compute the power spectrum. In the one-dimensional case the Fourier transform of $\xi(\tau)$, extended to negative τ as an even function of τ, is

$$P(k) = \frac{1}{a}\left(\frac{n_c}{n} - 1\right)\frac{2\sqrt{2}}{\pi}\frac{1}{k^2}\sin^2\frac{ka}{2}.$$

From the graph of this function it is clear that most of the power resides in Fourier components with wavelength $\lambda = 2\pi/k$ greater than a (figure 1.4).

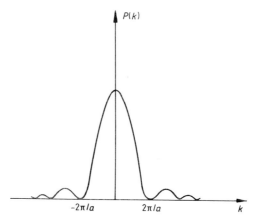

Figure 1.4. The form of the function $k^{-2}\sin^2 ka/2$.

Note, for later reference, that if $n_c/\bar{n} \gg 1$, we can neglect the edge effects calculated under (ii); these affect only the precise way in which the correlation function tends to zero near $\tau = a$.

1.7.2. A Power Law Distribution

Of interest subsequently will be a spherically symmetric distribution of density

$$n = n_0 r^{-\alpha}, \qquad \tfrac{3}{2} < \alpha < 3, \qquad r > r_0$$

in three dimensions. Since there are no statistical effects to worry about here, the computation of the correlation function is achieved by a straightforward integration. We have

$$\int_{\text{all space}} n(\mathbf{r})\, n(\mathbf{r}+\boldsymbol{\tau})\, \mathrm{d}^3\mathbf{r} = \iiint n_0^2 r^{-\alpha}|\mathbf{r}+\boldsymbol{\tau}|^{-\alpha/2} r^2 \sin\theta\; \mathrm{d}r\, \mathrm{d}\theta\, \mathrm{d}\phi$$

$$= \tau^{3-2\alpha} \iiint n_0^2 x^{-\alpha+2}(x^2+1-2x\cos\Theta)^{-\alpha/2}\sin\theta\; \mathrm{d}\theta\, \mathrm{d}\phi\, \mathrm{d}x$$

$$(1.7.1)$$

where $x = r/\tau$ and $\mathbf{r}\cdot\boldsymbol{\tau} = r/\tau\cos\Theta$. The integral clearly yields only a numerical factor which is of no interest here; we should however check that the resulting correlation

16

function is finite in the limit of infinite volume of integration. To see this note that

$$\bar{n} = \frac{1}{V} \int_V n(\mathbf{r}) \, \mathrm{d}^3 \mathbf{r} = \frac{3}{R^3} \int_0^R n_0 r^{-\alpha+2} \, \mathrm{d}r = \frac{3n_0 R^{-\alpha}}{(3-\alpha)},$$

and hence $\bar{n} \to 0$ as $R \to \infty$. Therefore $\bar{n}^2 V \sim$ constant $\times R^{3-2\alpha}$ as $R \to \infty$, which is the same dependence on R as the integral in (1.7.1), and therefore the ratio $\langle n(\mathbf{r}) n(\mathbf{r}+\tau) \rangle / \bar{n}^2$ is finite. From (1.6.3) we obtain a finite value for $\xi(\tau)$ of the form

$$\xi(\tau) \propto \tau^{3-2\alpha}.$$

Note that the power law density gives rise to a power law correlation function with no characteristic knee, and hence no preferred length scale (figure 1.5). The power spectrum in this case is easily obtained directly. We have, for the Fourier transform $\hat{n}(\mathbf{k})$ of $n(r)$,

$$\hat{n}(\mathbf{k}) = (2\pi)^{-3/2} n_0 \int r^{-\alpha} \exp(i\mathbf{k} \cdot \mathbf{r}) \, \mathrm{d}^3 \mathbf{r}$$

$$= (2\pi)^{-1/2} n_0 \iint r^{-\alpha} \exp(ikr \cos \theta) r^2 \sin \theta \, \mathrm{d}\theta \, \mathrm{d}r,$$

with $k = |\mathbf{k}|$, where the axis of polar coordinates has been chosen as the direction of $\mathbf{k}$. Hence

$$\hat{n}(\mathbf{k}) = k^{\alpha-3} (2\pi)^{-1/2} n_0 \iint x^{-\alpha} \exp(ix \cos \theta) x^2 \sin \theta \, \mathrm{d}\theta \, \mathrm{d}x.$$

Thus $P(\mathbf{k}) \propto \hat{n}(\mathbf{k}) \hat{n}^*(\mathbf{k}) \propto k^{2\alpha-6}$ and this, of course, shows no characteristic range of wavelengths.

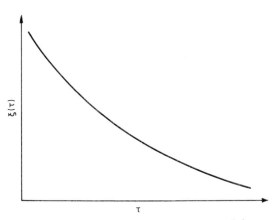

Figure 1.5. The form of the correlation function $\xi(\tau) \propto \tau^{3-2\alpha}$. This shows no characteristic scale of clustering.

1.7.3. Clusters of Various Sizes

We shall show in this example that the effect of adding clusters of various sizes in the right numbers leads to an approximate power law for the correlation function. This means that knowledge of the correlation function does not lead to a unique result for the distribution of galaxies. Of course, this is to be expected, since a complete knowledge of the distribution is equivalent to a knowledge of all the n-point correlation functions, and not just the two-point function.

Let there be $An^{-\beta}$ clusters containing n galaxies for each $n = 1, 2, 3, \ldots$, with A and β constants, and $2 < \beta < 3$. The total number of galaxies in volume V is $N = A \Sigma_n n^{-\beta+1}$ and the average density is $n_0 = N/V$. Let the clusters of n galaxies have radius r_n and volume V_n; the density in such a cluster is $\bar{n} = n/V_n$.

Retaining only the contributions from pairs of points lying in the same cluster, an approximation which was justified in the one-dimensional example (§1.7.1), we have

$$\langle n(\mathbf{r})\,n(\mathbf{r}+\boldsymbol{\tau})\rangle \sim \sum_{\substack{n \\ r_n > \tau}} \{\text{probability that } \mathbf{r} \text{ and } \mathbf{r}+\boldsymbol{\tau} \text{ lie in a cluster } C_n\} \times \{\text{value of the product in } C_n\}.$$

If we consider the case when one of the cluster members lies within $(r_n - \tau)$ of the cluster centre, the computation of the probability is straightforward; the converse case can be shown to give a somewhat smaller contribution to the average, although with the same final τ dependence (1.7.3), and we shall neglect it here. Our approximation for the average is then

$$\langle n(\mathbf{r})\,n(\mathbf{r}+\boldsymbol{\tau})\rangle \sim \sum_{\substack{n \\ r_n > \tau}} \bar{n}^2 \frac{(r_n - \tau)^3}{r_n^3} \frac{An^{-\beta}V_n}{V}. \tag{1.7.2}$$

Assuming that large clusters do not dominate, which is true if β is sufficiently large, we can extend the sums to infinity to obtain an approximate result. In addition we approximate the sums by integrals. Thus

$$N \sim A \int_1^\infty n^{1-\beta}\, dn = \frac{A}{\beta-2},$$

and

$$1/V = n_0/N = n_0(\beta - 2)/A,$$

provided $\beta > 2$. The sum in (1.7.2) becomes

$$\frac{3}{4\pi}(\beta-2)\frac{n_0}{A}\int_{r_n=\tau}^\infty n^{2-\beta} r_n^{-6}(r_n - \tau)^3\, dn.$$

The dependence of the cluster radius on the number of members has not so far been specified. Let us take $r_n = Bn^\alpha$, where B = constant and $\alpha > 0$. Setting $x = r_n/\tau$

as the independent integration variable, we obtain

$$n_0^2[1+\xi(\tau)] \sim \frac{3}{4\pi} \frac{(\beta-2)}{\alpha} \frac{A}{B} n_0 \tau^{-3+3/\alpha-\beta/\alpha} \int_1^\infty x^{-\beta/\alpha+3/\alpha-7}(x-1)^3\,dx.$$

If $4\alpha > 3 - \beta$ the main contribution to the integral comes from $x \gg 1$ (and hence from clusters with $r_n \gg \tau$), and we are justified in neglecting points further than $(r_n - \tau)$ from a cluster centre. Since $\alpha > 0$, this requires $\beta < 3$. Thus, for $\xi \gg 1$, and hence for small τ, we have

$$\xi(\tau) \propto \tau^{-3+3/\alpha-\beta/\alpha}, \tag{1.7.3}$$

which is the result we set out to establish.

1.8. The Angular Correlation Function

The catalogues of galaxies contain information on the positions of galaxies on the celestial sphere, but not on their distance. Therefore the correlation function $\xi(\tau)$ cannot be estimated directly from the catalogues. We define instead an angular correlation function $w(\theta)$ expressing the probability of finding pairs of galaxies separated by an angle θ in solid angles $d\Omega_1$, $d\Omega_2$ according to

$$dP_{12} \propto \bar{\sigma}^2 [1+w(\theta)]\,d\Omega_1\,d\Omega_2, \tag{1.8.1}$$

where $\bar{\sigma}$ is the average surface density of galaxies per unit solid angle on the sky. This definition assumes, of course, that the distribution is isotropic on a large scale. We assume also that there is no obscuration of more distant galaxies by intervening ones; for the depth of the present surveys this is easily satisfied.

In the case in which the galaxies can be taken to be continuously distributed with surface density $\sigma(\theta, \phi)$ per unit solid angle, a formula analogous to (1.6.3) holds:

$$\bar{\sigma}^2 [1+w(\theta)] = \frac{1}{4\pi} \int [\sigma(\Omega_1) - \bar{\sigma}] [\sigma(\Omega_2) - \bar{\sigma}]\,d\Omega,$$

where $\Omega_1 = (\theta_1, \phi_1)$ and $\Omega_2 = (\theta_2, \phi_2)$ are directions on the celestial sphere, and the integration is over all pairs of directions separated by an angle θ.

Clearly a knowledge of $\xi(\tau)$ should enable us to find $w(\theta)$. It is not clear that one can go uniquely in the opposite direction, but at least a relation between $\xi(\tau)$ and $w(\theta)$ would permit a test of any hypothesised distribution $\xi(\tau)$ against the data. To establish the desired relation we need to use information on the luminosity function for galaxies; this tells us the depth of a survey of given limiting apparent magnitude. For simplicity, assume initially that all galaxies have absolute magnitude M^*.

Consider first the relation of the surface density σ to the depth of the sample. If the limiting magnitude is m_*, galaxies to a distance D_* pc contribute to the

19

surface density, where

$$M^* - m_* = -5 \log D_* + 5.$$

The surface density is therefore

$$\bar{\sigma} = \tfrac{1}{3} D_*^3 n.$$

Note that we assume no intergalactic absorption, again in very good agreement with observations.

The joint probability dP_{12} is given by integrating the three-dimensional distribution out to the depth of the survey:

$$dP_{12} \propto n^2 \left\{ \int_0^{D_*} r_1^2 \, dr_1 \int_0^{D_*} r_2^2 \, dr_2 [1 + \xi(r_{12})] \right\} d\Omega_1 \, d\Omega_2,$$

where $r_{12}^2 = r_1^2 + r_2^2 - 2r_1 r_2 \cos\theta$. Comparing this with (1.8.1) we find

$$\mathscr{W}(\theta) = 9 D_*^{-6} \int_0^{D_*} dr_1 \int_0^{D_*} dr_2 \, r_1^2 r_2^2 \xi[(r_1^2 + r_2^2 - 2r_1 r_2 \cos\theta)^{1/2}],$$

which is the required relation. To extract information from it we make two approximations. (i) Only galaxies at small angular separation are significantly correlated; this enables us to put $\cos\theta \sim 1 - \tfrac{1}{2}\theta^2$ for small θ. (ii) ξ differs from zero significantly only for small r_{12}, and hence for $r_1 - r_2 \ll r_1 + r_2$. From this it follows that $4r_1 r_2 = (r_1 + r_2)^2 - (r_1 - r_2)^2 \approx (r_1 + r_2)^2$. It also follows that we can set the limits of integration to be infinite without serious error. Put

$$u = \frac{1}{2D_*} (r_1 + r_2); \qquad v = \frac{(r_1 - r_2)}{D_* \theta}.$$

Then

$$\mathscr{W}(\theta) = 9 \int_0^\infty du \int_0^\infty dv \, u^4 \xi[(v^2 + u^2)^{1/2} D_* \theta] \, \theta \tag{1.8.2}$$

is the required relation between $\mathscr{W}$ and ξ.

As an important example we can compute $\mathscr{W}(\theta)$ for a power law $\xi(\tau)$. If

$$\xi(\tau) = A\tau^{-\alpha}$$

then

$$\mathscr{W}(\theta) = \left\{ 9A \iint du \, dv \, u^4 (v^2 + u^2)^{-\alpha/2} D_*^{-\alpha} \right\} \theta^{-\alpha+1}. \tag{1.8.3}$$

Consequently a power law correlation in depth gives rise to a power law angular correlation function. The important point is that if the true clustering has no characteristic scale then the angular correlation exhibits no characteristic angular scale.

A further important conclusion can be drawn from (1.8.2) or (1.8.3). We observe first that $\mathcal{W}(\theta)$ can be written as

$$\mathcal{W}(\theta) = D_*^{-1} F(D_* \theta)$$

for an appropriate function F. Suppose then we have a second sample catalogue of depth $D_*' = x D_*$. For this sample we find

$$\mathcal{W}'(\theta) = D_*'^{-1} F(D_*' \theta) = x^{-1} D_*^{-1} F(D_* x \theta). \qquad (1.8.4)$$

Thus it should be possible to superpose plots of the two angular correlations if the angular scale (abscissa) for the second sample is expanded by a factor x and the scale of the correlation (the ordinate) diminished by a factor x. Equation (1.8.4) is referred to as a 'scaling' relation for the angular correlation function. The extent to which the scaling is satisfied in practice is a measure of the validity of the assumption of homogeneity. For example, suppose the clustering is indeed independent of location, and hence of depth, for sufficiently deep samples. Then by doubling the depth of a survey we should expect to halve the angular scale of correlations for the same linear scale; and we should expect to double the probability of chance coincidences, or, correspondingly, to halve the excess probability.

In practice it is necessary to drop the assumption that galaxies have the same absolute magnitude M^*, and hence to take into account the fact that galaxies of different luminosities are seen to different depths. Provided that the spatial correlations of galaxies are independent of their luminosities, this makes a difference of detail only, for the results we have obtained must hold for each infinitesimal range of intrinsic luminosities and hence, just as in our discussion of the number-magnitude relation (§1.2), they can be extended to all luminosities by summation (Davis 1976).

1.9. The Two-dimensional Power Spectrum

In order to discuss the results that have been obtained from analyses of galaxy catalogues we need to introduce also the two-dimensional power spectrum. Two different but essentially equivalent methods have been used. One method is to define a power spectrum of the angular correlation function on the celestial sphere; the power is given by replacing the Fourier transform used above by a transform appropriate to the spherical geometry, and employing spherical harmonics. This approach requires the introduction of a large amount of mathematical machinery, but has the advantage of analytical elegance, especially when applied to the whole sky. In practice, however, we do not deal with whole sky surveys, because of obscuration by the Galaxy if for no other reason. This leads to complexities of detail. In the second method, the positions of sources in the sky are mapped on to the plane by an equal-area projection. Standard Fourier analysis techniques can then be used, and there is no need to introduce the technology of spherical harmonics. We shall therefore describe in outline this second approach, and present the results as if they have been obtained in this way, while noting that, in fact, most of the analysis in the literature has been carried out using spherical transforms.

The ensuing discussion has two aims. First, the power spectrum can be used to assess the statistical significance of an apparent departure from a random distribution; thus, given a non-zero correlation function we can use the power spectrum to investigate how likely it is that this has arisen by chance in a sample from a truly random distribution. Secondly, given that the clustering effects are real, we can estimate the average number of sources per cluster directly from the power spectrum.

Consider a distribution of N sources in a finite rectangular area $(-X \leqslant x \leqslant X, -Y \leqslant y \leqslant Y)$ in a two-dimensional plane geometry. While the continuous-fluid picture is useful for exposition, it is inappropriate for the analysis of most, if not all, of the existing galaxy catalogues, so we return to a picture in which the sources are point objects at positions $(x_j, y_j), j = 1, \ldots, N$. A simple way to make this transition from a continuous distribution with density $\sigma(x, y)$ is to divide the plane into squares of area ΔA such that $\sigma(x, y) \Delta A = 1$ in each square. A point source is then located at the centre of each square. The Fourier transform of the density, a_{lm} (where l and m are integers), is given by

$$a_{lm} = \frac{4}{XY} \int_{-X}^{X} \int_{-Y}^{Y} \exp\left[2\pi i \left(\frac{lx}{X} + \frac{my}{Y}\right)\right] \sigma(x, y) \, dx \, dy. \tag{1.9.1}$$

This reduces to a sum over each square of $\sigma \Delta A$ ($= 1$) multiplied by the value of $\exp i \left[(lx/X) + (my/Y)\right]$ at the centre of that square, coordinates (x_j, y_j), where the source is located. Hence

$$a_{lm} = \sum_{j=1}^{N} \exp 2\pi i \left(\frac{lx_j}{X} + \frac{my_j}{Y}\right). \tag{1.9.2}$$

For notational convenience we write

$$\frac{2\pi x_j}{X} = u_j, \qquad \frac{2\pi y_j}{Y} = v_j,$$

and the expression for the Fourier coefficients (1.9.2) becomes

$$a_{lm} = \sum_{j=1}^{N} \exp i(lu_j + mv_j).$$

An alternative way of obtaining this result is to note that for point sources the density $\sigma(x, y)$ in (1.9.1) is a sum of Dirac δ-functions,

$$\sigma(x, y) = \sum_{j=1}^{N} \delta(x - x_j) \delta(y - y_j).$$

22

We imagine that our Universe is one of a large number in each of which the points (x_j, v_j) are chosen at random. The average value of a_{lm} over this large number of universes is

$$\langle a_{lm} \rangle = \sum_{j=1}^{N} [\langle \cos \theta_j \rangle + i \langle \sin \theta_j \rangle]$$

$$= 0, \qquad (1.9.3)$$

where $\theta_j = (lu_j + mv_j) \pmod{2\pi}$. This follows because $\cos \theta_j$ and $\sin \theta_j$ are equally likely to be positive or negative for a random set of θ_j.

The power spectrum depends on $|a_{lm}|^2$ (see equation 1.6.6), the average value of which is given by

$$\langle |a_{lm}|^2 \rangle = \sum_{j=1}^{N} \langle |\exp i(lu_j + mv_j)|^2 \rangle + \sum_{j \neq k} \langle \exp i[l(u_j - u_k) + m(v_j - v_k)] \rangle$$

$$= \sum_{j=1}^{N} 1 + \sum_{j \neq k} [\langle \cos \theta_{jk} \rangle + i \langle \sin \theta_{jk} \rangle]$$

$$= N, \qquad (1.9.4)$$

provided $(l, m) \neq (0, 0)$. Therefore, for a random distribution, $\langle |a_{lm}|^2 \rangle$ is independent of wavenumber (l, m), and the power in each (finite) wavelength is, on average, the same.

The object now is to calculate how likely it is that $|a_{lm}|^2$ should be found to differ from N in a given finite sample of a random population. To this end we map the random process under discussion on to one for which the statistics are known. Thus $\exp i(lu_j + mv_j)$, for each fixed (l, m), can be thought of as a step of unit length in the complex plane in a direction related to the position of the jth random point. As we sum over a random distribution of galaxies, the vector

$$a_{lm}^J = \sum_{j=1}^{J} \exp i(lu_j + mv_j) \qquad (1.9.5)$$

represents the end-point of a random walk in the complex plane after J steps. From (1.9.4) we expect that after N steps the average root mean square distance from the origin in many trials will be $\sqrt{N}$. For N sufficiently large, the probability distribution for the position of the end-point of a random walk is given by a solution of the diffusion equation in the plane, normalised to unit total probability. The diffusion coefficient is determined by the expected root mean square distance from the origin, and the probability distribution satisfying both (1.9.3) and (1.9.4) is

$$p(x, y)\, dx\, dy \sim (\tfrac{1}{4}\pi N) \exp -(x^2 + y^2)/2N\, dx\, dy.$$

This is the probability of finding the end-point in an area $dx\, dy$ centred on (x, y)

23

after N steps as $N \to \infty$. Consequently, the probability of finding $z_{lm}^2 = |a_{lm}|^2/N$ in the annulus between z_{lm} and $z_{lm} + dz_{lm}$ is

$$p(z_{lm}^2)\, dz_{lm}^2 = \tfrac{1}{2} \exp(-z_{lm}^2/2)\, z_{lm}\, dz_{lm}. \qquad (1.9.6)$$

The probability distribution given by (1.9.6) is called a χ^2 distribution with two degrees of freedom, χ_2^2. Therefore if a value of z_{lm}^2 is computed by putting the observed galaxy location (x_j, y_j) in (1.9.2), the probability that this value arises from a random distribution can be obtained from tables of χ_2^2. This tells us how likely it is that the observed galaxy distribution is random on the length scale $(X/l, Y/m)$.

Suppose that, in fact, the galaxies are in randomly distributed clusters of area A_c with N_c galaxies per cluster, and that within each cluster the galaxies are distributed at random. We can again calculate the expected value of z_{lm}^2. If $(X^2/l^2 + Y^2/m^2)^{1/2} \ll A_c^{1/2}$, i.e. if the wavelength of the component of the power spectrum under consideration is small compared to a cluster diameter, then the argument proceeds as before, since the sum (1.9.5) is again a sum over randomly distributed phases. However, if $(X^2/l^2 + Y^2/m^2)^{1/2} \gg A_c^{1/2}$, then galaxies within a cluster contribute to (1.9.5) with approximately the same phase, since the differences in their coordinates are small compared to the wavelength. Each cluster therefore contributes a vector of length approximately N_c in the complex plane, and the sum over clusters is a random walk from cluster to cluster with N/N_c steps of length N_c. In the limit $N \to \infty$, we expect a root mean square distance from the origin N_c times larger than in the unclustered case. Consequently, in the case that galaxies are in clusters of length scale $(X^2/l^2 + Y^2/m^2)^{1/2}$, we expect to find $N_c \sim |a_{lm}|^2/N - 1$. The probability distribution is

$$p(z_{lm}^2)\, dz_{lm}^2 = 1/2N_c \exp(-z_{lm}^2/2N_c)\, z_{lm}\, dz_{lm}, \qquad (1.9.7)$$

which is analogous to (1.9.6). Fitting the values of z_{lm} for all l, m to (1.9.7), as computed from the observed distribution of galaxies in the one Universe we have available, yields an estimate of N_c on the hypothesis that all the galaxies are clustered. For clusters of varying membership numbers this approach gives an estimate of the *average* number of galaxies per cluster.

1.10. Observational Results

1.10.1. *Clustering of Rich Clusters*

The power spectrum method has been used to test for superclustering among the rich clusters of the Abell catalogue (Hauser and Peebles 1973). It is necessary first to establish the existence of such clustering; subsequently the power spectrum and correlation function can be used to provide information on the scale and amount of the clustering.

To test for clustering, the distribution of the power spectrum z_{lm}^2, as obtained from (1.9.2) for the Abell catalogues, is compared to the exponential distribution expected for a random sample (1.9.6). The fit is found to be poor. If, however, comparison is made with a random distribution of clusters containing N_c sources per cluster (1.9.7), a good fit can be obtained for $N_c \approx 2$. This suggests clustering with an average of two sources per cluster. The test cannot distinguish the case in which all sources are contained in superclusters of two clusters, from that in which, say, about 90% of the sources are unclustered and the remaining 10% occur in clusters of about ten members. This latter case would give an average value

$$\langle z_{lm}^2 \rangle = 1 \times (1-f) + N_c f \sim 1.9,$$

where f is the fraction of sources clustered in numbers N_c, for those (l, m) values corresponding to the scale of the clustering.

Plots of the power against length scale again show departures from a random distribution on scales of a few degrees. Since the survey does not cover the whole sky, the z_{lm} are not independent; thus the structure on the intermediate scales ($l \sim 20$) could in principle be due to power leaking in from the obvious irregularities on very small scales. The fit to an exponential distribution suggests that this is not the case, and that the effect is real.

A plot of $\mathcal{W}(\theta)$ against θ again clearly shows weak clustering on scales of a few degrees. This is further confirmed by dividing the data in the Abell catalogue into the different distance classes and thereby using the catalogue to represent several

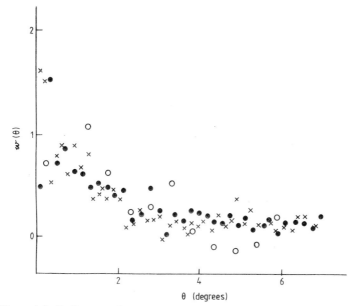

Figure 1.6. Scaling test for the covariance function for Abell clusters in the northern hemisphere. • w (θ) for distance classes 1–6; × w (θ) for distance classes 1–5 reduced to the same scale according to equation (1.8.4); ○ w (θ) for distance classes 1–4 reduced to the same scale (Hauser and Peebles 1973).

surveys to different limiting depths. We can then investigate whether the different surveys scale according to the theory of §1.8. The data for distance classes 1-4, 1-5 and 1-6 are shown reduced to the same scale in figure 1.6, and the agreement is clear evidence for the reality of the clustering. It also adds support to the assumption of overall homogeneity (§1.8).

1.10.2. Clustering of Galaxies

The catalogues of Zwicky, Shane and Wirtanen, and Rudnicki *et al* have been investigated by the correlation function technique. The angular correlation function is shown for each of the surveys in figure 1.7, and is reduced to the same scale according to the scaling laws of §1.8 in figure 1.8. Several features are immediately apparent.

(i) The correlation function goes to zero for large angles as required by the assumption of homogeneity. This result is essential for the application of the correlation function technique and validates our original assumption. If the correlation function did not tend to zero on large scales, then either the galaxy catalogues would not be a fair sample of a homogeneous Universe, or the Universe would not be homogeneous on any scale.

(ii) The scaling laws are well satisfied, showing that the information contained in the correlation functions represents a real physical clustering effect, and that the clustering is homogeneous on a large enough scale.

(iii) The data show no characteristic length scale. It can be fitted quite well by a power law $w(\theta) \propto \theta^{-0.77}$.

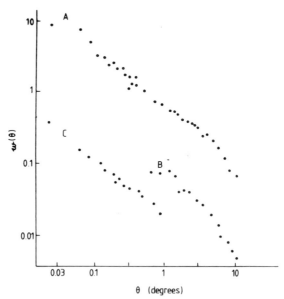

Figure 1.7. Correlation functions for the Zwicky (A), Shane–Wirtanen (B), and Jagiellonian (C) catalogues (Peebles 1975).

26

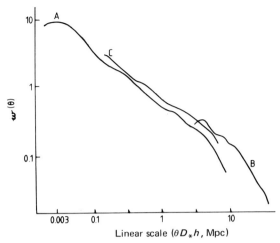

Figure 1.8. The data of figure 1.7 reduced to the same scale (Peebles 1975).

There is no *a priori* reason for choosing a power law in (iii), and other models are possible. For the power law model we obtain a spatial correlation function $\xi(\tau) = 20\tau^{-1.77}$ (§1.8). Thus at a distance of 1 Mpc from a given galaxy, we are 21 times more likely to find another galaxy than would be the case for a completely random distribution. This emphasises the fact that the clustering is quite pronounced on small scales.

The absence of a characteristic length in the clustering could be explained in several ways. The example in §1.7.3 shows that a random collection of clusters of all sizes approximates a power law provided there is a large excess of small clusters (e.g. $\alpha = 0.5$, $\beta = 2.4$). A more likely explanation, however, is that there is clustering of galaxies on all scales; smaller clusters are grouped into larger clusters with the amount of grouping together diminishing as we go to larger scales. This is a special sort of hierarchical picture; it is intermediate between the true hierarchical models, with an infinite hierarchy of equal degrees of clustering, and a random distribution, homogeneous on all scales, larger than some definite clustering scale. There appear to be no definite units out of which the large-scale homogeneous Universe can be taken to be constructed. Or, to put it less precisely, there is no largest type of object scattered homogeneously throughout the Universe of which the Universe can be thought of as being constructed, like the atoms of a gas.

A recent, more detailed reinvestigation of the Shane–Wirtanen catalogue has shown the existence of a sharp break in the angular correlation function at $\theta \sim 2.5°$, corresponding to a linear scale of ~ 9 Mpc (Davis *et al* 1977). The existence of this departure from a power law is important in the discussion of the evolution of galaxy corrections.

The three-point correlation function has also been estimated, and the result is consistent with

$$\xi(r_{12}, r_{23}, r_{31}) = \text{constant} \{\xi(r_{12})\,\xi(r_{23}) + \xi(r_{23})\,\xi(r_{31}) + \xi(r_{31})\,\xi(r_{12})\}$$

(Groth and Peebles 1977). This again is important information in any attempt to account for the development of correlations. It also shows that the three-point function does not introduce any length scales.

1.10.3. Cross Correlations

One would expect the superclustering tendency of clusters to be reflected in the clustering tendency of galaxies. Thus one would expect an excess over random of the probability of finding a galaxy in one catalogue near a cluster in Abell's catalogue. The expectation can be made quantitative in terms of a cross correlation function obtained by averaging the product of a pair of densities, one from each survey. The results are in good agreement (Seldner and Peebles 1977).

1.11. Classification of Normal Galaxies

So far we have investigated the cosmological information that can be derived from a consideration of galaxies as mere point sources of visible light. A detailed study of the various types of galaxies is not within the province of cosmology, and all that is really needed in the sequel is some knowledge of the existence of such objects as radio galaxies and quasars. Nevertheless, some of the points in the following discussion are of importance in, for example, the theory of galaxy formation.

We first distinguish between normal galaxies and active galaxies. As with all the distinctions in this subject there is no sharp dividing line between categories, the true distribution of galaxy types being essentially continuous. However, as long as this is recognised, a simple discrete classification scheme is very useful. In addition, the classification scheme usually includes objects such as quasars which are not even known to be galaxies in the sense of aggregates of stars. Broadly speaking a normal galaxy is one in which the radiation comes mainly from normal stars, albeit possibly modified by dust, etc. In general, in an active galaxy stars play at most a minor role as energy sources, the dominant contribution to the radiation being from non-thermal sources in the nucleus or, as in the case of many radio galaxies, in lobes well separated from the parent galaxy. There may be exceptions to this since in some active galaxies, the dominant radiation source could be an abnormally large number of hot OB stars. The different types of active galaxies will be dealt with in §1.12.

Since we do not yet understand how galaxies form and what physical parameters or initial conditions or evolutionary processes determine the differences between galaxies, the classification schemes are observationally based. There are therefore, in principle, as many classification schemes as there are combinations of measurable parameters, most of which, it sometimes seems, have appeared in the literature.

28

The most prevalent classification scheme derives from Lundmark and Hubble and is based on the morphology of the galaxies. The scheme is summarised in Hubble's famous 'tuning fork' diagram (figure 1.9).

Figure 1.9. The Hubble classification of galaxies.

The ellipticals show no substructure apart from a strong concentration in the nucleus. The surface brightness $I(r)$ of the brighter ellipticals $(-23 \lesssim M_v \lesssim -16)$ falls off with radius as

$$I(r) = I_0 \exp\left(-\alpha r^{1/4}\right)$$

away from the nucleus and outer edge. For the dwarf ellipticals $(-16 \lesssim M_v \lesssim -9.5)$ the surface brightness decreases more rapidly towards the outer parts. Ellipticals are generally flattened systems and the observed ratio of axes b/a, which depends both on the intrinsic flattening and unknown inclination, is used to subdivide the class. A class En elliptical has $b/a = (1-n/10)$. Observed values of n range from 0 to 7, so $b/a \gtrsim 0.3$.

The lenticular galaxies, denoted S0, are like ellipticals in their absence of substructure, but resemble spirals in their intrinsically disc-like structure. Typically, $b/a \sim 0.25$ for both S0s and spirals. The average surface brightness distributions for both classes are also similar, with an inner spheroidal component following the

29

distribution for an elliptical galaxy, and an outer exponential disc with

$$I(r) = I_0 \exp(-\beta r),$$

which contributes the major part of the luminosity.

Spirals resemble lenticular galaxies except in the additional features of the spiral arms and the presence of young stars and dust, and are divided into subclasses according to the nature of the spiral arms. Hubble recognised two families: the normal spirals in which the spiral arms emanate from the nucleus; and the barred spirals in which the arms begin at the ends of a bar. Each of the families is further subdivided into types a, b and c, along which sequence the size of the nuclear bulge decreases, and the spiral arms become relatively more unwound. About 3% of galaxies fall outside this classification scheme and are designated as irregulars.

The *Hubble Atlas of Galaxies* (Sandage 1961) gives an extensive catalogue of types based on a modified Hubble classification. The relative frequencies of the main classes are approximately $E:SO:S:Irreg = 13:22:62:3$.

Hubble's classification scheme has been modified and extended in many ways. For example, Van den Bergh (1959, 1960) has noted that the Hubble classification is usually made by reference to dwarf galaxies. This is indicated by an additional parameter to denote luminosity class in analogy with stellar luminosity classes.

A somewhat different approach is taken in the Yerkes classification described by Morgan (1958, 1959). This is based on a correlation of form with the integrated spectrum of galaxies. For example, those dominated by bright central nuclei have spectra corresponding to K giants. Several additional morphological types are introduced, some of which are commonly used in certain branches of astronomy. In particular, D systems are dustless, showing no pronounced structure; the most luminous, denoted cD, are often the dominant members of rich clusters – the cD notation has generally been adopted in radio astronomy. N galaxies are those containing small bright nuclei emitting up to 50% of the total light of the galaxy, and are commonly referred to in optical observations.

1.12. Active Galaxies

The basic feature of an active galaxy is a large luminosity not due to a 'normal' population of stars; in some cases it may arise from a large excess of hot stars, while in others from non-thermal processes such as synchrotron emission by relativistic electrons in a magnetic field – in principle it is possible that an active 'galaxy' may contain no stars at all. Since the physical principles underlying this activity are not understood, the classification systems depend on overlapping observational criteria.

Searches for active galaxies involve the observation of excess continuum emission in various parts of the electromagnetic spectrum or the detection of emission lines. For example, the surveys by Markarian (Markarian and Lipovetsky

1974) were based on searches for excess emission in the near-ultraviolet, and the Tololo survey (Smith 1978) identified objects with strong emission lines. Radio surveys have revealed the existence of radio galaxies and quasi-stellar sources (radio quasars). Many active galaxies are most luminous in the infrared, and x-ray emission appears to be a common feature of quasars, type 1 Seyferts and BL Lac objects (*v.i.*).

The most active of active galaxies are the quasars; at the distances implied by their redshifts, quasar luminosities range from 10^{39} to 10^{41} W. Their optical spectra show broad emission lines (up to 30 000 km s^{-1}) superimposed on a power law continuum typically $i_\nu \propto \nu^{-1}$. This is similar to type 1 Seyfert galaxies and, indeed, the most luminous Seyferts, removed to greater distances, would be indistinguishable from less energetic quasars. Type 2 Seyferts have only narrow lines, with widths of the order of 10^3 km s^{-1}. A few narrow-line quasars are also known; a catalogue of Seyfert galaxies is given by Weedman (1977), and catalogues of quasars by De Veny *et al* (1971) and Burbidge *et al* (1977).

BL Lac objects resemble quasars but show at most only weak emission lines and are further characterised by high polarisation and rapid variability. Optical luminosities are of the order of 4×10^{38} W. Closely related to Seyferts, but having in addition significant radio fluxes, are the broad- and narrow-line radio galaxies. Where the underlying galaxies can be classified, Seyferts are almost without exception found to be spirals, while the emission line radio galaxies are associated with ellipticals.

Radio galaxies in general are taken to be those associated with sources of luminosity greater than 10^{33} W from normal galaxies; typical values are in the range 10^{35}-10^{38} W. Major catalogues of the radio sky, particularly for the purpose of optical identification, are the Cambridge 3C catalogue for the northern hemisphere and the Parkes PKS survey for the southern hemisphere. References to the major catalogues are given in Moffett (1975) and in Rees *et al* (1974). Sources are often double — two expanding regions of radio emission distributed about the central galaxy — or triple, as in the case when the central galaxy is also a strong source. Multiple sources sometimes occur. Dumb-bell galaxies (two nuclei with a common envelope) are always single. Strong radio emission (10^{37}-10^{39} W) is associated with about 5% of quasars, again of the double or triple type.

1.13. Classification of Clusters

For some purposes rich clusters of galaxies can be usefully classified into different types. The details of various classification schemes are of no importance in this book, so we merely note their basic characteristics. Observational properties which have formed the basis of classification schemes are: morphology, the dominance of the brightest members, and the distribution of galaxy types. To a large extent these schemes correlate with each other and with other cluster properties (Bahcall 1977).

The classification according to morphology is based essentially on the amount of central concentration of member galaxies and the degree of spherical symmetry of the cluster. This ranges from regular clusters of the Coma type, with over 1000 members highly centrally condensed, to irregular clusters, such as Virgo, with no central concentration and no symmetry. The Rood–Sastry (RS) scheme is a refinement of this in that the arrangement of the brightest members is taken into account. For example, Coma is of binary type since it is dominated by two bright galaxies.

The Bautz–Morgan (BM) classification has achieved a certain popularity amongst radio astronomers. It relates to the brightness of the dominant galaxy (or galaxies) relative to the other members. A type I cluster is dominated by a central cD galaxy; type II by galaxies intermediate between cD and normal giant ellipticals; and type III contains no dominant galaxies. In this scheme Coma is type II and Virgo is type III.

The regular clusters are composed dominantly of elliptical galaxies, especially in the central core, whereas irregular clusters contain a relatively large proportion of spirals. Thus the morphological scheme correlates with a classification based on galactic content.

1.14. Clustering of Types of Galaxy

From the discussion of the galactic content of rich clusters of galaxies in §1.13, we might expect that the clustering tendency of galaxies will vary with type; in particular, that the amount of clustering should diminish along the sequence ellipticals, lenticulars, spirals. This is confirmed, at least approximately, by analysis of the two-point correlation functions relating to the individual types. Amongst active galaxies we find that Seyferts, being spirals, tend not to occur in clusters. Although individual cases of quasars associated with clusters are known, there is no evidence for more than a random association. On the other hand, strong radio sources tend to be associated with the brightest elliptical cluster galaxies, and one might expect this to be apparent in catalogues of radio galaxies. Indeed, the cross correlation between the 4C radio catalogue and the Shane–Wirtanen galaxy counts follows a power law of the same form as the galaxy correlation function (§1.10.2), but with a larger coefficient. This can be interpreted to mean that radio sources are associated with the most strongly clustered galaxies (Seldner and Peebles 1977).

The radio data are particularly important since they can be used to probe the Universe more easily to a greater depth than with optical observations. This arises because the radio sources, at least at high Galactic latitudes, can be assumed to be extragalactic, since there are few discrete Galactic sources of any significance out of the Galactic plane. This means that the mere detection of a radio source provides cosmological information and it is not necessary to obtain and identify a galaxy image as in optical work.

Early analyses of the radio data suggested that in some surveys anisotropies in the distribution and pronounced clustering were present, but this was contradicted by analyses of other surveys. More recent work provides firm evidence for the large-scale homogeneity of the Universe.

Webster (1976) has used a version of the method of power spectra to test for weak clustering primarily on angular scales $\gtrsim 4°$. Rather than the z_{lm}^2 themselves, Webster uses the running average

$$Q_{\lambda'}' = \frac{1}{\nu} \sum z_{lm}^2 ,$$

where the sum is over l, m such that the wavelengths λ' of the terms included lie in some convenient range $\lambda^{-1} < (\lambda')^{-1} < \lambda^{-1} + \Delta$, and ν is the number of terms in the sum. For a completely random distribution $Q_\lambda' = 1$, and for a distribution with an average of N_c sources per cluster $Q_\lambda' = N_c$ when the wavelength λ equals the cluster scale.

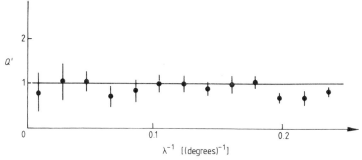

Figure 1.10. Plot of Q' against Y_λ for a $75° \times 75°$ area of sky centred on $\alpha = 4^h$, $\delta = +31.5°$ for the 4C survey with 1σ error bars (Webster 1976).

Figure 1.10 shows a graph of Q' against angle (or λ^{-1}) for a 'square' of side $75°$ centred on RA $= 4^h$, declination $-6° < \delta < 69°$ from the 4C survey. There is no evidence that Q' is greater than unity, and hence there is no evidence for clustering. The slightly negative values of Q' are explained by source confusion where close pairs of sources are unresolved.

This work surveys a portion of the Universe out to 5000 Mpc, much further than is reached in optical investigations. Most of the sources in the radio catalogues are at large distances. Consequently, the result again demonstrates forcefully the large-scale homogeneity and isotropy of the Universe.

2. The Expanding Universe

2.1. Olbers' Paradox

The most surprising aspect of the night sky, once one has absorbed the presence of the stars, is the existence of the dark spaces between the stars. The paradoxical nature of this darkness, first pointed out by Halley and Le Chésaux, has come to be known as *Olbers' paradox*. To state the problem we can assume the Universe to be uniformly populated by sources, which we should nowadays think of as galaxies or clusters rather than stars, each having luminosity L. If these sources have number density n, the flux at the Earth is

$$F = \int \frac{nL}{4\pi r^2} \cdot 4\pi r^2 \, dr. \tag{2.1.1}$$

This means that in an infinite universe, in which the laws of Euclidean geometry are valid, the night sky is infinitely bright. In this calculation we have neglected the fact that our line of sight to the more distant galaxies will be obscured by foreground galaxies. Taking this into account we can argue that, since our line of sight intersects a star at some point, the average sky brightness must be the same as the average surface brightness of a star. This scarcely goes any way to removing the contradiction. Nor would the presence of obscuring material since this would simply be heated to the stellar surface temperature.

The paradox can be resolved if we can restrict the range of integration in equation (2.1.1). To achieve this by simply asserting that the Universe is finite in space or in time is somewhat unnaturally *ad hoc*. A more natural solution is a universe in which the sources are expanding away from some common origin a finite time in the past. The range of integration in (2.1.1) is then the light travel time over the age of the Universe, and is therefore finite. Another less important effect is that the light emitted by the receding galaxies is Doppler-shifted to the red end of the spectrum, and is thus weakened in energy; this reduces the flux from each source. On the other hand, the overall expansion means that at earlier times the sources we are seeing now were part of a more compressed Universe, so n was higher. This effect does not however spoil the finiteness of (2.1.1). In Chapter 3 we shall, in fact, use the measured finite value of the brightness of the night sky to set limits on the number of galaxies in the visible Universe which are too faint to be seen as discrete sources.

We do not argue that the expansion of the Universe provides the only possible resolution of Olbers' paradox, but nevertheless this is the way it is resolved in our Universe. Conversely, in a contracting universe, the sky would grow increasingly bright. We cannot say that the darkness of the night sky proves that the Universe is expanding; it is, however, the most obvious manifestation of that expansion.

2.2. The Redshift

The wavelengths of the spectral lines we observe from an individual star in the Galaxy do not correspond exactly to the wavelengths of those same lines in the laboratory. The lines will be shifted to the red or blue by an amount depending on the velocity of the observed star relative to the Earth. The overall relative velocity is the sum of the rotational velocity of the Earth, the velocity of the Earth round the Sun and the solar system around the Galaxy, as well as any peculiar motion of the star. The order of magnitude of this effect is given by the rotational velocity of the Galaxy, since this is the largest contribution. At the solar system this is about 250 km s^{-1}. The first-order Doppler shift in wavelength, $\Delta\lambda = (\lambda_0 - \lambda_e)$ for a velocity v is given by

$$\Delta\lambda/\lambda_e = v/c.$$

The redshift, z, is then *defined* by

$$z = \Delta\lambda/\lambda_e,$$

and hence the ratio of observed to emitted wavelength is

$$\lambda_0/\lambda_e = 1 + z.$$

Negative z corresponds, of course, to a blue shift. We therefore expect for the Ca II line at 3969 Å a shift of order ± 4 Å corresponding to $z \approx \pm 10^{-3}$, which is much larger than the width of the lines due to motion in the stellar photosphere. Note that, in terms of frequencies, we have $\nu_0/\nu_e = (1+z)^{-1} \approx 1 - z$ for small velocities, and hence that $z \approx -\Delta\nu/\nu_e$. Of course, these formulae are valid only if $v \ll c$.

If we look at an external galaxy we see not individual stars but the integrated light of many stars. The spectral lines will therefore be *broadened* by $\Delta\lambda/\lambda \sim 10^{-3}$ because of the motion of the stars in that galaxy. For the members of the Local Group, typical velocities are of the order of a few hundred km s^{-1}, giving rise to red or blue shifts of the same order as the line broadening. For some of the brightest nearby galaxies we find velocities ranging from 70 km s^{-1} towards us, to 2600 km s^{-1} away from us, but as we go to fainter and more distant galaxies, the shifts due to these peculiar velocities become negligible compared with a systematic redshift. If, in this preliminary discussion, we interpret this systematic effect as a Doppler velocity of recession, this leads to a picture of an overall expansion of the system of galaxies upon which are superimposed relatively small peculiar motions.

The Doppler interpretation of the redshifts is in agreement with the results from radio observations of the 21 cm emission line of atomic hydrogen in external galaxies. The velocities computed from these observations agree with optical values over the range $-300\,\mathrm{km\,s^{-1}}$ to $4000\,\mathrm{km\,s^{-1}}$, as expected for the Doppler effect. Nevertheless, one should be aware that the naive interpretation of the redshift as a straightforward Doppler effect, convenient as it is for an initial orientation, is by no means an accurate picture. This will become apparent in Chapters 5 and 6.

2.3. The Expanding Universe

In view of the isotropy and homogeneity of the distribution of matter discussed in Chapter 1, it is natural to assume that the rate of recession of the galaxies from us is the same in all directions. The isotropy of expansion is logically distinct from the isotropy of the distribution even if, from our present naive viewpoint, it is difficult to imagine how a direction in which galaxies were receding more slowly could avoid having an excess of brighter (i.e. nearer) galaxies. We shall also assume that we are not situated at a privileged central location, and that the expansion appears isotropic to any observer. At any one time, therefore, the rate of expansion must be the same at any point in space, and so the expansion is homogeneous. This conclusion is clearly related to, but again is logically distinct from, the result that a distribution of matter which is isotropic about each point must be homogeneous.

With these assumptions we find that the expansion of the Universe can be described by a single function of time, $R(t)$, called the *scale factor*. Furthermore, the assumptions lead to a definite prediction for the dependence of the recessional velocity of a galaxy on its distance. Note that we neglect here the clustering of galaxies and assume an exactly homogeneous distribution. The connection with observations will be made more precisely below.

Consider the three galaxies shown in figure 2.1, at times t_0 and t, and look at the expansion from the point of view of observer A. Isotropy implies that the increase

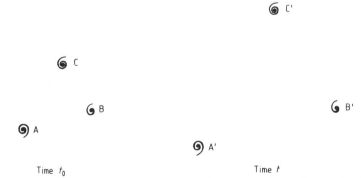

Figure 2.1. Three galaxies at A, B and C at time t_0 expand away from each other to A′, B′ and C′ at time t.

36

in distance $AB \rightarrow A'B'$ be the same as $AC \rightarrow A'C'$, but from the point of view of the observer at C isotropy requires that the expansion in AC be the same as in BC. The sides of the triangle are expanded by the same factor, say $R(t)/R(t_0)$. Adding a galaxy out of the plane to extend the argument to three dimensions, we see that isotropy about each point implies that the expansion is controlled by a single scale factor, $R(t)$, giving the ratios of corresponding lengths at different times. If AB has length l_0 at time t_0 and $l(t)$ at time t, we have

$$l(t) = l_0 R(t)/R(t_0).$$

Therefore the relative velocity, $v(t) = \mathrm{d}l(t)/\mathrm{d}t$, is given by

$$v(t) = \frac{1}{R(t_0)} \frac{\mathrm{d}R(t)}{\mathrm{d}t} l_0 = \frac{1}{R(t)} \frac{\mathrm{d}R(t)}{\mathrm{d}t} l. \tag{2.3.1}$$

For small velocities we can use the non-relativistic Doppler formula to obtain for the observed redshift of a galaxy at present distance $l(t)$,

$$cz(t) = l(t)\, \dot{R}/R, \tag{2.3.2}$$

where we have written $\dot{R}$ for $\mathrm{d}R/\mathrm{d}t$. Therefore, on the basis of homogeneity and isotropy alone, we have arrived at a linear relation between the redshift of a galaxy and its distance, at least for sufficiently nearby galaxies.

In order to make a meaningful comparison with observation, it is necessary to decide what exactly it is in our not exactly uniform Universe that is supposed to be expanding according to (2.3.2). Clearly, such objects as atoms, the Earth, the Sun and the Galaxy do not expand because they are held together as bound systems by internal electrical or gravitational forces. Consider, for example, the Galaxy. The gravitational potential, $(GM/r)_{\mathrm{Galaxy}}$, measures the strength of the internal binding and gives a dimensionless escape velocity, $v_{\mathrm{esc}}/c = (2GM/rc^2)^{1/2}$, of the order of $v_{\mathrm{esc}}/c \sim 10^{-3}$. The recessional velocity of the edge of the Galaxy as seen from its centre would be given by (2.3.1) as $v/c = (\dot{R}/R)r/c$. For $\dot{R}/R = 10^2\,\mathrm{km\,s^{-1}\,Mpc^{-1}}$ (see §2.4), this is $v/c \sim 3 \times 10^{-6}$, which is negligible compared to the escape velocity, so the internal gravitational forces dominate over the expansion of the Universe. A typical rich cluster has a mass 10^2–10^3 times a galactic mass, and a radius 2×10^2 times the radius of the Galaxy, and this again leads to a bound system. For nearest-neighbour clusters, however, taking the intercluster distance to be about three times the cluster diameter, we obtain $v_{\mathrm{esc}}/c \sim 5 \times 10^{-4}$, whereas $(\dot{R}/R)r/c \sim 2 \times 10^{-3}$. Thus separate clusters are typically not bound together by gravity. Consequently, to a first approximation we should regard the clusters of galaxies, rather than the galaxies themselves, as the basic units or particles of an expanding Universe.

To investigate the validity of the linear relation (2.3.2) we therefore look at the correlation of the mean redshift of a cluster with its distance. To a good approximation we can take the apparent brightness of the brightest galaxy in a cluster as a measure of distance. This assumes that the absolute magnitude of the brightest galaxy is constant from cluster to cluster. This can be verified for clusters sufficiently near to have their distances determined by other means (§2.4), and there is no evidence in cluster morphology for any systematic evolution. There may be some correlation between the absolute magnitude of the brightest galaxy and cluster richness, but this is a small effect.

Figure 2.2, taken from Sandage (1972) and Kristian *et al* (1978), shows a recent plot of redshift against apparent magnitude after corrections for galactic absorption. There is also a *K*-correction; this arises from the fact that apparent magnitudes are measured for a fixed range of wavelengths at the receiver, but for galaxies of different redshifts this must correspond to different ranges of emitted wavelengths. The *K*-correction allows for the wavelength dependence of the intensity of the radiation emitted by galaxies. For nearby galaxies the *K*-correction can nowadays be eliminated by the use of multi-channel photometry, which allows the direct comparison of intensities at a fixed rest frequency.

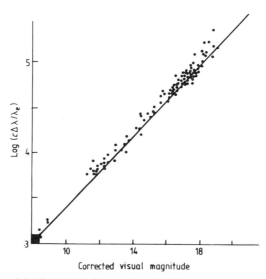

Figure 2.2. The Hubble diagram for the brightest galaxies in clusters.

The small black rectangle in figure 2.2 represents the data available to Hubble in his pioneering investigation, and from which he discovered the linear relation between redshift and distance now known as Hubble's law. In terms of apparent magnitude this is

$$\log z = 0.2\,m + \text{constant.} \tag{2.3.3}$$

This prediction made on the basis of homogeneity and isotropy, is clearly confirmed over a wide range of reshifts, but we have no reason here to expect Hubble's law to hold exactly for arbitrarily large redshifts.

2.4. The Distance Scale

The principle of the method by which the distances to clusters of galaxies are measured is as simple as its detailed implementation is complicated. To start with we need one absolute measurement which is nowadays provided by radar ranging determinations of the scale of the solar system. Determinations of the parallax of nearby stars yields the ratio of their distances to the radius of the Earth's orbit. This method can be used to about 50 pc. With some exceptions, the subsequent steps in the distance scale out to the limits of the observed Universe then follow a standard pattern; a property of sources at known distances is found which is reasonably intrinsically constant from source to source; it is then assumed that this property has the same intrinsic value in more distant sources, so that its measured apparent value can be related to the distance.

Various types of properties can be used. The luminosity of a particular class of source is an obvious candidate. Examples of this are stars of given spectral type, as in the method of spectroscopic parallax, and stars with a characteristic luminosity-dependent variability, such as the pulsations of Cepheids and the light curves of supernovae. A variant of this method is the use of the brightest object of a given class, such as the brightest globular clusters, or the brightest galaxy in a cluster, with the assumption that the intrinsic scatter in luminosity will be small for a sufficiently large sample. Of course, consideration must be given to evolutionary effects which may give rise to systematic variations in absolute luminosity with cosmic epoch, and this is indeed a problem in the measurement of the largest distances. An alternative property which scales conveniently with distance is the linear size of an object of given type. This has been applied to the sizes of galactic H II regions and the separation of the lobes of radio emission from radio galaxies and quasars. A summary of the cosmic distance ladder is sketched in figure 2.3.

A complementary approach to distance determination is based on the fact that distance is the ratio of a linear scale to an angular scale. Distances can be measured to objects for which a linear and angular size (or a linear and angular velocity) can be determined independently. The method is therefore 'absolute' in the sense that it is independent of other rungs on the distance ladder.

An example of this scheme is the determination of the distance to nearby galactic clusters. The proper motion of the cluster members appears to converge to a point on the celestial sphere at which their parallel velocities meet. This gives the direction of motion of the cluster, θ, to the line of sight. From Doppler measurements of the radial velocities, v_r, we get the transverse velocities of cluster members, $v = v_r \tan \theta$. The proper motion, angular velocity μ, then gives the distance of the cluster, $d = v/\mu$.

A second important example is the supernova method (Kirschner and Kwan 1974). The linear size of the shell of expanding gas in a supernova outburst is estimated from observations of the velocity of the shell and the time elapsed by assuming a uniform rate of expansion. The angular extent is derived from the ratio

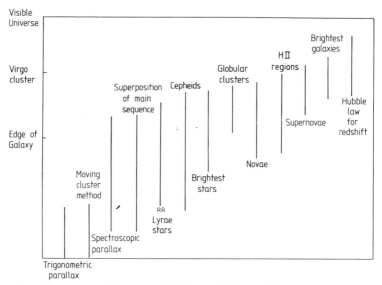

Figure 2.3. Some of the rungs of the cosmic distance ladder. The vertical scale gives an approximate indication of a logarithmic scale of the distances over which the methods labelled can be used.

of the flux received to that radiated by a black body at the spectroscopically determined temperature of the emitting gas. Observations of supernovae in external galaxies give their distances directly, and the results agree broadly with other methods.

2.5. The Hubble Constant

Once a distance scale has been established we can convert from apparent magnitude to distance in the Hubble relation (2.3.3). Thus we write

$$z = H_0 d/c, \qquad (2.5.1)$$

where H_0 is called the Hubble constant, and d is the distance. From equation (2.3.2) we can identify H_0 with the current value of the Hubble parameter

$$H = \dot{R}/R.$$

Note that the observable quantity H_0 is independent of the fiducial length l_0 in equation (2.3.1); this must be true for all observable quantities since the choice of l_0 is quite arbitrary. Thus the scale factor itself cannot be measured, but only ratios of scale factors.

The isotropy of the expansion is expressed through the independence of H_0 on direction. Any anisotropy of the expansion would appear as a departure ΔH from the mean Hubble constant $\bar{H}_0$. Thus $\Delta H/\bar{H}_0$ is a measure of the isotropy of the expansion. This is known to be less than 30%, so that direct observation of the Hubble constant provides this weak limit for the possible anisotropy of the Universe.

40

Hubble's original estimate of H_0 was $500 \, \text{km s}^{-1} \, \text{Mpc}^{-1}$; as is well known, this is too high. There were two reasons for this high value: one was the uncertainty of the correction for galactic absorption in determining the calibration of the period–luminosity relation for the Cepheid variables, and the other was the misidentification of H II regions in external galaxies as bright stars. Taken together, these factors lead to a reduction of H_0 to about $100 \, \text{km s}^{-1} \, \text{Mpc}^{-1}$. Recently, Sandage and Tammann (1975) have reinvestigated the distance scale and claimed a reduction in H_0 by a further factor of about 2. The work is exceedingly difficult and this value has not received universal acceptance. In particular, de Vaucouleurs (de Vaucouleurs and Bollinger 1979) has argued for a Hubble constant rather nearer $100 \, \text{km s}^{-1} \, \text{Mpc}^{-1}$. In view of these uncertainties we shall adopt the usual expedient of writing $H_0 = 100h$ and keeping track of h in all formulae so that they can be adjusted as required. Unfortunately, some authors adopt $H_0 = 50h$, so care is required in reading the literature.

2.6. The Age of the Universe

The Hubble parameter defines the rate at which the clusters of galaxies are moving away from each other. If we assume for the moment that the velocity of expansion is constant, we find that all of the clusters must have emanated from a mathematical point at a finite time in the past. Since two galaxies separated by a distance d_0 move apart with velocity $v = Hd = H_0 d_0$, the present age of the Universe, under the assumption of a constant expansion rate, is H_0^{-1}. Note that, by homogeneity, no one point in space can be regarded as the centre of the expansion; rather, all points in space came together at the initial time. Hubble's value of the age (H_0^{-1}) is $1/500 \, \text{km}^{-1} \, \text{s Mpc}$, or, in more suitable units for this purpose, $1/500 \times 3 \times 10^{19} \, \text{s} \approx 2 \times 10^9$ years.

In fact, the Hubble parameter probably does not obligingly change with time in such a way as to keep the recessional velocities constant. Intuitively one might expect the rate of expansion of the Universe to be slowing down as it gets older because of the gravitational attraction of matter. The precise rate at which it does so is measured by the deceleration parameter q, defined by

$$q = -\frac{\ddot{R}R}{\dot{R}^2}. \tag{2.6.1}$$

The deceleration parameter is essentially just the second derivative of the scale factor, but is constructed in such a way as to depend only on ratios of scale factors and to be independent of the units of time. Observations appear to support the intuitive notion that q_0, the present value of q, should be positive but this may be open to dispute (see Chapter 7). Thus the velocity of recession was probably greater in the past, and the Universe has taken less time than H_0^{-1} to expand to its present state. Therefore the age of 2×10^9 years would seem to represent an upper limit. This is unfortunate since the Sun is 4.5×10^9 years old and the Earth is certainly older than 2×10^9 years.

The resolution of this paradox is now thought to rest simply on the revision of the value of the Hubble constant. For $H_0 = 100 \, \text{km s}^{-1} \, \text{Mpc}^{-1}$ we have $H_0^{-1} \approx 2 \times 10^{10}$ years, which is a reasonable upper limit. However, this apparent conflict in the original expanding universe models was of historical importance since it led directly to the Steady State Theory.

2.7. The Steady State Theory

Once the evolution and hence finite lifetime of astrophysical systems within the Universe is accepted (in particular the irreversible processing of material in the evolution of stars), it becomes difficult to see how the Universe could be infinitely old. The expanding Universe of finite age represents one way out of the problem. The alternative is a regenerating mechanism, that is, the creation of new matter. We do not observe the creation of matter in the laboratory, so there is no empirical or theoretical foundation for such an hypothesis. The best one can do is to try to make the simplest assumption and see where it leads. This is the assumption that the regeneration of the Universe occurs at just such a rate as to keep its average properties constant in time. The Universe is expanding, since this is required by observation, but matter is created to keep the density constant. Indeed, in this vein, all physically observable properties of the Universe must be constant on average. In particular, H must be constant; consequently

$$\mathrm{d}l/\mathrm{d}t = H_0 l,$$

and so

$$l = \exp(H_0 t) \, l_0.$$

The expansion is therefore exponential, with $R(t) = \exp(H_0 t)$, and not linear as one might have guessed. The exponential curve is, of course, self-similar in the sense that increasing t by a given amount is merely equivalent to rescaling l_0. There is therefore no privileged origin for time. From (2.6.1) we obtain $q = -1$.

The Steady State Theory can be subjected to observational tests both of its basic non-evolutionary philosophy and in its detailed prediction of a constant H. It fails both of these tests — it fails to account for the evolution of radio sources (Chapter 7) and quasars (§2.8), and it is ruled out by the limits on q_0 obtained from the detailed shape of the redshift–distance relation at large redshifts. Perhaps the most serious difficulty is its failure to account in any natural way for the microwave background (Chapter 3). The Steady State Theory is now therefore of historical interest only. In particular, it does not satisfy our criterion of taking known physics to explore the Universe (§5.1).

2.8. The Evolving Universe

If we could look back at the Universe to a redshift of about 300 we should see no galaxies at all. This is because at about this redshift the galaxies would begin to overlap. Thus galaxies must have formed at or after this stage, and at earlier times there were no galaxies. The properties of galaxies themselves must change with time. Consider, for example, the distribution of quasars on the assumption that the observed redshifts of quasars may be taken to indicate their distances according to Hubble's law (2.5.1). The sample of optically identified quasars is believed to be complete down to optical magnitude 18.5. For each quasar in the sample the distance is known from its redshift, z, so the redshift out to which it could be placed and still be bright enough to appear in the sample, z_m, can be calculated. We can then estimate the average value of the ratio, V/V_m, of the volume, V, within which each quasar is found, to the maximum volume V_m within which it could be found. If quasars are uniformly distributed in an expanding Universe it can be shown that $V/V_m = 0.5$ (Schmidt 1968). The observed value is 0.7, which indicates a higher spatial density of bright quasars in the past, over and above the effects of expansion. This can be accounted for either by assuming that quasars turn off after a finite lifetime, or that they grow dimmer with time. It is clear that the expanding Universe presents us with the problem of the evolution of structure (Chapter 12).

Another aspect of this evolution is presented by the discovery of the microwave background radiation (Chapter 3). This has a natural explanation in a hot big-bang model of the Universe; not only was the Universe denser in the past than it is now, but it was also hotter. To understand this, and to develop its consequences, means that we must study the evolution of interacting matter and radiation in an expanding system (Chapter 8).

Not only is our evolution from the past of interest, but so too is the course of our future evolution. The question as to whether the Universe will go on expanding forever or will eventually halt and collapse has always been a central issue of cosmology. On the basis of current theory, and with the help of observations, we can make certain predictions. Of crucial importance here is the determination of the present density of matter in the Universe which is taken up in Chapter 4.

3. The Quality of Radiation

3.1. Cosmic Radiation

The Universe lives through the conversion of matter into energy. This is most obvious in the light of individual stars, but these stars shine against a background of radiation from those distant galaxies too faint to be seen as individual sources. We may therefore distinguish between discrete emission from identified sources, and a background or diffuse radiation field — a distinction that depends, of course, on the sensitivity of our instrumental eyes.

Outside the optical band of the spectrum there is discrete and diffuse emission of a largely non-stellar origin. At the radio end of the spectrum the major contribution to the background is provided by diffuse emission from the Galaxy. An average value for the magnetic field in the Galaxy is 2×10^{-10} T, as determined by observations of Faraday rotation and dispersion in radio emission from pulsars. Moving in this field are high-energy cosmic ray electrons, with energies exceeding 10 J per electron, produced, presumably, in supernovae. This results in the emission of synchrotron radiation at radio frequencies; indeed, the discovery of this radiation constituted the first extension of astronomy beyond the optical window. From independent data on the energy spectrum of the cosmic ray electrons (§4.14) one can deduce a spectrum for this Galactic radio emission, and the smaller extragalactic contribution to the radio background can then be determined. A power law spectrum for the intensity, or surface brightness $i(\nu) \propto \nu^{-\alpha}$, is obtained with α estimated to be between 0.7 and 0.9; the measured intensity at 178 MHz is about 1.6×10^{-22} W m^{-2} sr^{-1} Hz^{-1}. This flux is most simply explained as the integrated emission of sources too weak or too distant to be seen individually: known extragalactic discrete sources, such as radio galaxies and quasars, have appropriate spectra, and plausible extrapolation from the observed brighter sources yields the correct integrated intensity. The energy density of the radiation is about 10^{-20} J m^{-3} in the band $10^6 \lesssim \nu \lesssim 10^9$ Hz and this is distributed throughout space. For comparison, note that the energy densities of the Galactic magnetic field and of cosmic rays in the Galaxy are both of order 10^{-13} J m^{-3}.

In contrast, there are, as far as we know, no sources emitting significantly in the microwave region, where thermal radiation would correspond to a temperature of a few degrees kelvin. Nevertheless, a relatively intense flux of radiation ($\sim 10^{-5}\,\mathrm{W\,m^{-2}}$) is observed, with properties that make it difficult to hypothesise any reasonable local emitters. It therefore appears that the radiation must be propagating to us from cosmological distances. It was the discovery of this radiation that stimulated the renaissance of physical cosmology in the 1960s, and led to the emergence of cosmology from the vapours of speculation as a branch of physical science.

One expects the situation for the infrared background to parallel that at radio frequencies. Re-radiation of absorbed starlight by interstellar dust grains should provide a Galactic infrared background and active galaxies should yield the main extragalactic contribution, although no definitive results have yet been obtained in this spectral range.

At optical wavelengths only upper limits to the extragalactic background are available since this is dominated by Galactic starlight. However, as discussed in Chapter 4, these limits are important in restricting the possible numbers of faint galaxies. The surface brightness is less than about $5 \times 10^{-23}\,\mathrm{W\,m^{-2}\,sr^{-1}\,Hz^{-1}}$ at $6 \times 10^{14}\,\mathrm{Hz}$ (~ 5000 Å). For comparison, the energy density, u, in the optical band of Galactic starlight is $10^{-13}\,\mathrm{J\,m^{-3}}$ giving an intensity, or surface brightness, $i_\nu \sim uc/4\pi\nu \sim 2 \times 10^{-21}\,\mathrm{W\,m^{-2}\,sr^{-1}\,Hz^{-1}}$.

Beyond 912 Å to about 100 Å, in the far-ultraviolet, the interstellar medium itself is opaque, since photons of these wavelengths are absorbed by photoionisation of hydrogen. In the near-ultraviolet around $10^{15}\,\mathrm{Hz}$ the background is less than $10^{-24}\,\mathrm{W\,m^{-2}\,sr^{-1}\,Hz^{-1}}$.

At the high-energy end of the electromagnetic spectrum the x-ray background has been measured between 0.25 keV (≈ 100 Å) and 1 MeV. The origin of the radiation is not known; it is probably the integrated emission from active galaxies and quasars, but diffuse emission from intergalactic space could provide a significant contribution (§3.5). Observations of gamma rays, conventionally regarded as photons with energies greater than 0.5 MeV (the rest mass energy of an electron), yield upper limits for the background flux. These are important, for example, in discussions of the antimatter content of the Universe (§4.15).

The electromagnetic spectrum does not exhaust the possibilities for storing energy in the Universe in the form of radiation, by which we mean particles of zero rest mass. We must also consider neutrino waves and gravitational waves, although as yet neither has been detected in a cosmological context. Neutrinos are characteristically different from photons and gravitons since they have half-integral spin and so obey Fermi-Dirac statistics at low temperatures. Since they interact only weakly with other particles there is no way of keeping them hot against the cooling by expansion in the expanding Universe (§8.4). Indeed, a background of both neutrinos and antineutrinos could exist in the present Universe since the rate of mutual annihilation is so slow. Any neutrinos (or antineutrinos) will therefore be degenerate and fill up the available energy states up to the Fermi level $E_F = (ch)^{3/4}(4I/c)^{1/4}$ for a neutrino surface brightness I. The best available estimate of the maximum allowable surface brightness is $I \lesssim 2 \times 10^7\,\mathrm{W\,m^{-2}\,sr^{-1}}$, corresponding to a Fermi level of 2 eV. A background density of neutrinos with greater energies would not be possible since interactions with high-energy protons would prevent these from travelling a significant distance across the Galaxy, in contradiction to the observations of cosmic rays. This limit, however, is very weak, and is equivalent to at least 10^5 times as much energy in neutrinos as in the rest mass of galaxies. Nevertheless, it is an order of magnitude or so better than the limit obtained from laboratory experiments on beta decay. If the recently reported observation of a non-zero rest mass for neutrinos (Reines *et al* 1980, Lyubimov 1980) is correct, then neutrinos would contribute significantly to the density of matter in the Universe and could play an important role in the initiation of galaxy formation at early times (see Chapter 12).

Although gravitational waves have not yet been detected directly, their existence is fairly certain since the radiation of gravitational energy appears to be in quantitative agreement with the orbital evolution of the binary pulsar PSR 1913+16 (Hulse and Taylor 1975, Taylor *et al* 1979). The standard picture of the expanding Universe (§8.5) requires the existence of gravitons with a number density of at least the same order of magnitude as the microwave photons ($\sim 10^8$ m^{-3}) with a black-body spectrum peaking at around microwave frequencies (10^{11} Hz). The detection of such a weak flux is quite beyond the realms of possibility. We expect events such as supernovae and galaxy formation to contribute to a gravitational wave background, the former at frequencies around 1 kHz (corresponding to a time-scale of a few milliseconds for the collapse of the stellar core), and the latter at around 10^{-14} Hz (corresponding to a wavelength of 1 Mpc typical of the separation of galaxies). It is easy to see, without calculation, that such a background is unlikely to be detectable directly, for we should not expect these events to produce significantly more energy in gravitational waves than in electromagnetic radiation. However, because the gravitational interaction is so much weaker, for detectability we would require many orders of magnitude more flux in gravitational waves. Only if there is a large density of gravitational radiation not associated with an electromagnetic background, such as may be produced, for example, in a very inhomogeneous early stage of the Universe, would background gravitational waves be observable. The same argument applies of course, *mutatis mutandis,* to the integrated emission of neutrinos.

We shall consider the microwave and x-ray backgrounds in turn in the following sections. At present, by far the most important background for cosmology is the microwave radiation. According to the simplest picture we have of the origin of this radiation, it is not a product of the conversion of matter to energy. If we omit from our discussion the first 10^{-6} seconds of the history of the Universe, so that we confine ourselves to conditions in which the physical processes are understood, then we can regard the microwave photons as part of the initial input into the Universe. This gives us the 'hot big-bang' model of the Universe which has proved so successful in accommodating the observations in a coherent picture. On this view the microwaves are signals from 'creation' and much of our subsequent discussion will be concerned with the information that they bring us. The hot big-bang interpretation of the microwave background is not necessarily the only possible model for the rarities of Nature's truth, but at the very least it provides a fruitful working hypothesis.

3.2. The Microwave Background

Down to wavelengths of about 3 mm there are atmospheric windows which permit ground-based observations. However, at wavelengths shorter than about 50 cm the Galactic emission is negligible and is thought not to rise again until one reaches the radiation from dust in the far infrared at around 0.04 cm. Nor are the known discrete extragalactic sources expected to give any contribution in this microwave region.

It therefore came as something of a surprise to Penzias and Wilson at the Bell Telephone Laboratories to find an apparently non-terrestrial source of radiation at 7 cm interfering with their communication satellite experiments (Penzias and Wilson 1965). This was less of a surprise some 30 miles down the road in Princeton, where, at the same time, a radiometer was under construction to look for just such a signal (Dicke *et al* 1965). For Dicke had independently rediscovered an argument, due originally to Gamow, that the temperatures and densities required in the early Universe for the synthesis of helium from hydrogen would be remembered in a ubiquitous background of microwave radiation at the present day. This is discussed further in Chapter 8; for a somewhat less condensed history see Peebles (1971 and references therein) and Weinberg (1977).

The early results on the spectrum of the microwave radiation are consistent with a black-body distribution at a temperature of about 3 K. The intensity of such radiation peaks around 0.2 cm which is beyond the window of atmsopheric transmission (figure 3.1). Ground-based observations can therefore be made only on the

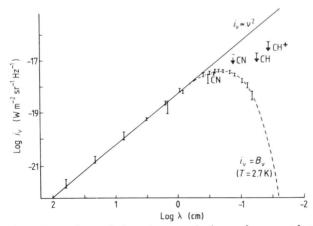

Figure 3.1. Observations of the microwave background compared with the Rayleigh–Jeans and black-body curves at 2.7 K.

Rayleigh–Jeans part of the Planck spectrum, where, to a good approximation, the intensity i_ν at frequency ν is given by

$$i_\nu = 2\nu^2 kT/c^2.$$

Observations of i_ν yield a temperature T only on the basis of the *assumption* of a thermal spectrum.

Some confirmation that the spectrum turns over as required came from the observation of certain molecular absorption lines. Many molecules can exist in rotationally excited states, the lower levels of which have excitation energies, ΔE, corresponding to wavelengths near the supposed black-body peak at 0.2 cm. Bathed in thermal radiation at 3 K, a large fraction of such molecules will exist in their first excited states. Some of these molecules, such as cyanogen (CN), can give absorption lines in the *optical* band, which can therefore be observed from the ground. Comparison of the strengths of the absorption by CN from the ground state at $\lambda = 3874.608$ Å, and from the first excited state at $\lambda = 3873.998$ Å, as observed in the optical spectra of stars seen through absorbing clouds, yields the relative populations of the states. The Boltzmann formula for the level populations

$$n_1/n_0 = g_1/g_0 \exp(-\Delta E/kT),$$

where the ratio of statistical weights $g_1/g_0 = 3$ for the states under discussion, then yields a temperature T for the exciting radiation. The failure to observe absorption from the second excited state of CN, or from the first excited states of CH and CH$^+$ radicals, provides upper limits to the temperature of the radiation. For a truly thermal spectrum, of course, the temperature translates into an intensity at the wavelength $hc/\Delta E$ corresponding to the excitation energy, and this is plotted in figure 3.1. The observations, made for many stars, yield results consistent with a universal background radiation.

Early attempts to measure the radiation field beyond the peak directly from above the atmosphere in balloon-borne and rocket experiments gave discrepant results. The thermal character of the background has, however, been confirmed by Woody *et al* (1975) with measurements from a balloon, as indicated in figure 3.1 (see also Robson *et al* 1974). The temperature was found to be 2.9 K to within about $\pm 5\%$. More recent data (Woody and Richards 1979) appear to show a small but significant excess near the peak, which could cause serious difficulties for the hot big-bang model to be discussed in this book.

That this radiation is non-local in origin is shown by its isotropy on large angular scales. Thus no association is found, for example, with the Galactic plane, the Local Group or the Virgo supercluster. In fact, as we shall see, the large-scale isotropy is sufficiently exact to be used to provide information on the large-scale structure of the Universe with much greater precision than can be obtained from the counts of galaxies considered in Chapter 1. Quantitative results for the isotropy are most easily expressed in terms of temperature fluctuations associated with the measured intensity at various wavelengths through the Planck formula, which allows the results to be characterised by a single number. The isotropy of the temperature can be determined with much greater precision than the absolute value of the temperature itself. Thus, a typical upper limit at various wavelengths and on angular scales of 360° and 180° is

$$\delta T/T \lesssim 10^{-3}.$$

Positive detections of a 360° variation have been reported by several investigators (see §3.4). This can be interpreted in terms of our random motion relative to the overall Hubble expansion (§3.4). Subtracting this from the data leaves a residual $\delta T/T \lesssim 3 \times 10^{-4}$ on large scales.

Measurements at various wavelengths on small angular scales ranging from about 1° to a few minutes of arc again yield typically $\delta T/T \lesssim 10^{-3}$. The strictest limit claimed is $\delta T/T < 3 \times 10^{-5}$ over 3.6′ (Parijskij 1973) and $\delta T/T \lesssim 8 \times 10^{-5}$ over 7′ (Partridge 1980). There are also claims that positive detections of fluctuations in the background temperature have been made in the direction of the Coma cluster; this is discussed further in §9.5.

From the temperature of the background radiation we deduce an energy density $u = aT^4 \approx 0.4 \times 10^{-13}$ J m^{-3}, comparable with starlight within the Galaxy, but spread, of course, throughout a much larger volume.

3.3. The Hot Big Bang

From the observation that the microwave background is approximately isotropic, it follows that the source of the emission cannot be anything in our local environment up to the scale of, say, the Local Supercluster. In fact it would appear that we have been saved a certain amount of work by having been born away from the centre of the Supercluster, for we should not otherwise have been so easily able to rule out the possibility of microwave emission from Supercluster material. It now follows that the background must be universal. Suppose, on the contrary, that the background were associated with some larger scale of clustering. Within such a region we see galaxies with large recessional velocities. We must therefore ask how a background which appears isotropic to us would appear from such galaxies. Looking away from us, observers in these galaxies would see radiation Doppler-shifted to higher frequencies; looking towards us they would see radiation Doppler-shifted to the red. We shall show later that the shifts will be such as to retain a black-body spectrum in each direction, but having a temperature that varies across the sky. For the moment the precise form is unimportant; the relevant point is that the background would appear highly anisotropic to these observers. It would then follow that the isotropy of the microwave radiation is for us an accident of our location in the Universe. Since we have no independent evidence for this, we can regard it as sufficiently unlikely to merit no further consideration. We conclude that the background must be a universal constituent of the expanding Universe.

We can now use our knowledge of the detailed small-scale isotropy to rule out the possibility that the microwaves arise from a universal population of unresolved discrete sources. Now, if one is allowed a complete freedom of *ad hoc* hypotheses, then a discrete-source model is evidently possible. For example, if the sources are supposed to be distributed in a uniform lattice over the sky there is no possibility of conflict with observations of isotropy for suitably chosen lattice spacings. This, however, is clearly a grossly unrealistic distribution. At the very best one might argue for a random distribution, although even then it is unreasonable to suppose the sources can entirely avoid the non-random clustering of galaxies. However, we do not have to discuss this objection, for we can show that even a random distribution would show a clumpiness that can be ruled out by the data.

Let n be the mean number of sources in the beam of a radio telescope of given aperture. For a random distribution the probability of m sources appearing in the beam is given by a Poisson distribution

$$p(m) = \frac{n^m e^{-n}}{m!}.$$

The most favourable hypothesis for the discrete-source model is that each source has the same apparent luminosity, since this gives the minimum fluctuation. The mean square variation from the average (the expected value of $(m - n)^2$) then gives the square of the expected fluctuation in the intensity of the microwave signal as the telescope is moved across the sky. This is

$$\sum_m p(m)(m - n)^2 = \sum_m m^2 p(m) - 2n \sum_m m p(m) + n^2 \sum_m p(m).$$

We have

$$\sum_m p(m) = 1, \qquad \sum_m m p(m) = n,$$

by definition, and

$$\sum_m m^2 p(m) = \sum \frac{m^2 n^m e^{-n}}{m!} = e^{-n} n^2 \sum_{m>2} \frac{n^{m-2}}{(m-2)!} + e^{-n} n \sum_{m>1} \frac{n^{m-1}}{(m-1)!}$$

$$= n^2 + n.$$

Therefore

$$\frac{\delta i}{i} = \frac{\sqrt{n}}{n} = \frac{1}{\sqrt{n}}, \tag{3.3.1}$$

which is just the usual $n^{-1/2}$ fluctuation of a random distribution. On the Rayleigh–Jeans tail of the Planck spectrum we have $\delta i/i = \delta T/T$, and so from the observed $\delta T/T \lesssim 0.001$ it follows that $n \gtrsim 10^6$.

A typical angular scale is, say, $10'$, corresponding to a solid angle of 7×10^{-6} sr. From this we calculate a source density of about 10^{11} sr^{-1} for typical observed values, not the best available. This is possibly an order of magnitude greater than the number of galaxies in the visible Universe, and we have made no allowance for the inverse square dilution of flux from the more distant sources! Thus it would appear that we can maintain a discrete-source model only by populating the Universe with an absurd abundance of unclustered sources, which reveal their presence in no other way.

One might attempt a discrete-source model by arguing that the Universe could be relatively opaque to the microwave background and therefore that radiation from individual sources could be smoothed out by scattering. The problem in this case is that there is no evidence for scattering in the high-redshift objects studied at neighbouring wavelengths in the radio and infrared or optical bands. Consequently, it would also be necessary to postulate an as yet unknown source of opacity which manifests itself only in the microwave region.

A further alternative to the cosmological origin of the microwave radiation which has been proposed is diffuse thermal re-radiation from intergalactic grains or 'dust' heated by the absorption of radiation from galaxies. Again, given suitable freedom in the choice of shape, size and composition of the grains, it is possible to account for both the isotropy and spectrum of the background within the limits of observational error. For example, cylindrical needles of graphite avoid the unobserved extinction in the optical which would be produced by otherwise suitably sized spherical grains, and should give an approximate black-body spectrum over at least a large part of the waveband. Although further observations are required to rule out such models, they are usually considered to be rather *ad hoc* and contrived compared with the hot big-bang hypothesis.

There is an alternative version of this theory which has recently received some attention (Rees 1978). This involves the hypothesis of a pregalactic epoch of star formation. The energy involved could be reprocessed in the 'interstellar' medium at large redshifts ($z \gtrsim 100$) to yield the observed microwave background. In this picture the microwave background would not provide us with direct information on the initial stages of the Universe. Whether it is a viable alternative to the hot big-bang model depends on the prediction and observation of detailed departures from a thermal spectrum.

To see the simplest way of understanding the existence of a universal radiation field we emphasise again the thermal character of the background. The only straightforward way to produce a thermal spectrum is to leave matter and radiation together undisturbed for long enough for them to achieve thermal equilibrium. It is clear that the contemporary Universe is not in thermal equilibrium, since not everything in our environment is even approximately at the 3 K of the background radiation. One can think of this as another aspect of Olbers' paradox: the expansion of the Universe means in effect that the matter and radiation are not undisturbed. We shall show in §§3.6 and 8.1 that, once produced, a thermal spectrum of radiation remains thermal as the Universe expands, although the temperature of the radiation decreases. This suggests we should look to some earlier phase of the Universe to provide the conditions of thermal equilibrium. It also tells us that in the past the background radiation must have been hotter. We know already that in the past the matter must have been denser. At sufficiently high temperature and density thermal equilibrium is achieved rapidly (§8.4); it is therefore natural to postulate an early stage in the history of the Universe at which matter and radiation were in thermal equilibrium at high temperature and density. We can then follow the development of such a system as it expands and cools to compare it with the Universe we know now. This is the hot big-bang model. The initial stage may be taken back as far as it is believed that one is dealing with temperatures and densities at which our laws of physics can be applied. This is at least as early as a Universe only about 10^{-5} seconds old, at which the density is of the order of the densities to be found inside the nuclei of atoms. With a suitable dose of arrogance or scepticism to taste one can extrapolate the known laws further back to 10^{-44} s, at which time it appears that a so far undiscovered quantum theory of gravity would be essential. Before this is the as yet unknown hot big bang.

3.4. *Eppur si muove* (Nevertheless it Moves)

Once we believe that we know what the microwave background is, its properties provide information on the nature of the Universe, a theme to which we shall find several occasions to return. Here we discuss how a positive detection of anisotropy in the background (Smoot *et al* 1977) has been used to determine the motion of the Earth (thereby confirming Galileo's comment, which provides the title of this section, on his at the time much publicised recantation).

To understand the result we first have to study how a thermal radiation field which is isotropic to one observer appears to another in uniform relative motion with velocity v. We need the result only to order v/c, so we can ignore the special relativistic effects of time dilation and length contraction, which are of order $(v/c)^2$, and use Newtonian mechanics. To discuss the specific intensity of radiation in a direction at an angle θ to the motion of the observer, imagine a telescope pointing in this direction and receiving radiation from a small solid angle about this direction. As far as the telescope is concerned it is moving with velocity $v \cos \theta$ through a uniform density radiation field, from which it extracts radiation incident in a small solid angle about the forward direction. The intensity of radiation received is the energy per unit frequency interval crossing unit area of telescope mirror per unit time divided by the solid angle subtended. We want to relate this to the analogous quantity measured by a stationary telescope. There are three differences between the moving and stationary systems. First, the moving telescope collects more flux simply by virtue of the fact that it is sweeping up photons: in time dt, the number of photons passing the mirror is increased by a factor $(c\, dt + v \cos \theta\, dt)/c\, dt = 1 + \beta \cos \theta$, where $\beta = v/c$. Secondly, the energy of each photon received is increased by the

Doppler effect by a factor $(1 + \beta \cos\theta)$, since the energy is proportional to frequency; however, the flux is also spread over a larger band width, $d\nu' = (1 + \beta \cos\theta)\,d\nu$ and the net result for the specific intensity, which is the energy per unit band width, is that the two effects cancel. Thirdly, the solid angle for the moving telescope is smaller by a factor $(1 + \beta \cos\theta)^{-2}$ as a result of aberration. Thus, to first order in β, the intensity in direction θ for the moving observer, $i_{\nu'}'(\theta)$, is related to the isotropic intensity i_ν in the rest frame by

$$i_{\nu'}'(\theta) = (1 + \beta \cos\theta)^3 i_\nu, \qquad (3.4.1)$$

where

$$\nu' = (1 + \beta \cos\theta)\,\nu.$$

Now let us see the effect on a Planck spectrum,

$$i_\nu = (2h/c^2)\,\nu^3\,[\exp(h\nu/kT) - 1]^{-1}.$$

We have, from (3.4.1),

$$i_{\nu'}'(\theta) = (2h/c^2)\,\nu'^3\,[\exp\{h\nu'/k(1 + \beta \cos\theta)\,T\} - 1]^{-1},$$

which is just a black-body distribution with temperature

$$T' = T(1 + \beta \cos\theta).$$

That the Planck spectrum in particular should remain unchanged in form reflects the fact that a system in thermal equilibrium must appear so for all inertial observers.

Thus the microwave sky should appear hottest in the direction of motion and coolest in the opposite direction with a 'dipole' variation of the form

$$\delta T/T = v/c \, \cos\theta.$$

Observations of $\delta T/T$ can therefore be used to find the velocity of the observer, v.

At the very least, the Earth cannot be at rest relative to the background radiation at all seasons of the year. In fact, the annual motion around the Sun contributes only a relatively modest 30 km s^{-1} to our progress through the Universe. Of much greater importance are the motion of the Sun around the Galaxy, which amounts to some 300 km s^{-1}, the motion of the Galaxy in the Local Group, which is possibly of the same order of magnitude, and the motion of the Local Group in the Virgo supercluster, which is essentially unknown but possibly small. It is highly improbable that these motions in unrelated and changing directions should cancel, so we expect an anisotropy of the order of $\delta T/T \sim v/c \sim 10^{-3}$, which is close to the earlier observed upper limits.

Because of the rotation of the Earth, such an anisotropy should appear in a fixed radiometer as a signal variation with a period of one sidereal day (which is just the time taken for the telescope to return to point towards a direction in the sky fixed relative to the stars, not the Sun). This enables the required signal to be extracted from both the noise and any other real effects in the data. Recent results (Smoot *et al* 1977, Muller 1978, Cheng *et al* 1979, Smoot and Lubin 1979, Fabbri *et al* 1980) yield a velocity of $390(\pm 30)$ km s^{-1} in the direction RA = 11^h, $\delta = +6°$. Figure 3.2 shows that a large peculiar velocity of the Galaxy is required to produce the observed result; if the velocity of the Local Group relative to Virgo is indeed small this leads to the somewhat surprising result that the supercluster as a whole must have a substantial peculiar velocity.

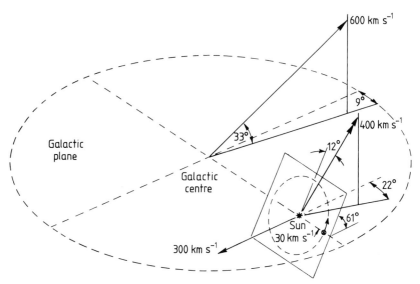

Figure 3.2. The motion of the Earth (from Muller 1978). The velocity measured relative to the microwave background is shown by the double arrow.

The frame of reference in which the microwave background appears isotropic can be regarded as providing a standard of absolute rest. This sometimes seems to lead to concern that there might be a conflict with the special theory of relativity, since this is held to assert the impossibility of establishing a privileged rest frame, even by using experiments involving electromagnetic radiation. That this cannot be the import of the principle of relativity is readily seen by considering the rumpus that would doubtless ensue were transportation companies to charge their passengers independently of whether they were in fact transported anywhere, on the grounds that all motion is relative. What relativity in fact forbids is the determination of motion by local experiments, experiments, that is, like the Michelson–Morley experiment, which can be performed (at least in principle) in laboratories shielded from external influences, for example by drawing the curtains. The non-local anisotropy observations measure our velocity relative to the microwave background and this is, in principle, no different from observing our motion relative to the stars.

3.5. The X-ray Background

The x-ray regime (3×10^{16}–3×10^{20} Hz) is the only region of the spectrum outside the microwave band where the diffuse background dominates the Galactic sources. Even so, the energy density is relatively small – a factor of about 10^{-2} less than the microwave background. Like the microwave radiation, the diffuse x-rays are isotropic on both large and small angular scales, and therefore cannot have a local origin. Over scales of $360°$ and $180°$, corresponding to sidereal periods of 24 h and 12 h, variations in intensity are less than about 0.2% (Warwick *et al* 1980); on scales of a few degrees one finds upper limits to the fluctuations of typically a few per cent (Schwarz 1970). The small-scale constraints in particular are not stringent enough to rule out the possibility that the x-ray background arises as the integrated emission of discrete sources, provided these are reasonably numerous and not too clustered. The spectrum of the radiation is definitely non-thermal, being best fitted by a power law with intensity i_ν given by

$$i_\nu = 3 \times 10^{-15} \, \nu^{-0.4} \, \text{W m}^{-2} \, \text{sr}^{-1} \, \text{Hz}^{-1} \tag{3.5.1}$$

from about 3×10^{17} to 5×10^{18} Hz (1–20 keV) and becoming somewhat steeper at higher energies (Schwarz 1970; see also Marshall *et al* 1980).

There is no agreed explanation for the main contribution to the x-ray background, although several processes which must each provide at least some of it are known. A number of Seyfert galaxies have been shown to be strong x-ray sources, some giving as much power in x-rays as in the optical, and the evidence is consistent with the assumption that strong x-ray emission is typical. Many quasars also are now known to be x-ray sources. With a modest amount of hypothetical evolution in the number density or luminosity of these sources, it is possible to account for the spectrum and intensity of the background as the summed emission from Seyferts and quasars which have not been individually identified. We must therefore consider whether these sources violate the small-scale isotropy constraints.

First we argue as for the microwave background in §3.3. Using, for example, the result that on a $5°$ angular scale, fluctuations in intensity are less than 5% (Fabian and Sanford 1971), equation (3.3.1) yields $n \gtrsim 400$ in 6×10^{-3} sr, or a source density in excess of 10^5 sr^{-1}. This is some three orders of magnitude less than the number of Seyferts. However, the simple theory of §3.3 makes no allowance for the inverse square dilution of radiation, which will mean that the fluctuations are dominated by the nearby brighter sources.

To estimate this effect, let the galaxies have x-ray luminosity L and consider them to be randomly distributed, with average number density n_0, in a sphere of radius c/H_0 which is to represent the observable Universe. The effects of expansion, such as the redshift of the incident energy, will be neglected. Then the average intensity di due to a shell of radius r, thickness dr, is

$$di = \frac{L}{16\pi^2 r^2} \langle dm(r) \rangle = \frac{L}{16\pi^2 r^2} \, dn(r), \tag{3.5.2}$$

where $dm(r)$ is the number of sources in the shell that happen to lie in the beam, and $\langle dm(r) \rangle = d \langle m(r) \rangle = dn(r)$ is the average number. For each shell we know from our previous calculation that the fluctuation in the square of the intensity due to the shell is proportional to the average number of sources in the shell that lie in the solid angle of the beam. Therefore,

$$d(\delta i)^2 = \frac{1}{(4\pi)^2} \left(\frac{L}{4\pi r^2} \right)^2 dn(r).$$

Since

$$dn(r) = n_0 \Omega r^2 dr, \tag{3.5.3}$$

where Ω is the solid angle of the beam, we find

$$d(\delta i)^2 = \left(\frac{L}{4\pi} \right)^2 \frac{\Omega}{(4\pi)^2} n_0 \frac{dr}{r^2}. \tag{3.5.4}$$

Now the contributions from each shell are independent since the distribution of sources is assumed to be random. Therefore, to obtain the total fluctuation we simply add the contribution from each shell. Furthermore, there will be a minimum radius below which individual sources can be identified and would not be deemed to be contributing to the diffuse background. Let this radius be r_m. Then, by integrating equations (3.5.4) and (3.5.2), using (3.5.3), we obtain

$$\left(\frac{\delta i}{i} \right)^2 = \frac{H_0^2}{n_0 \Omega c^2 r_m} \left(1 - \frac{H_0 r_m}{c} \right)^{-1}.$$

This is to be compared with (3.3.1) which, in this model, is equivalent to

$$\left(\frac{\delta i}{i} \right)^2 = \frac{1}{n} = \frac{3 H_0^3}{4\pi c^3 n_0 (\Omega/4\pi)}.$$

If $r_m \ll c/H_0$, a condition which is amply fulfilled for the x-ray Seyferts, we see that the effect of the inverse square dilution is to increase the fluctuations by a factor $[(c/H_0)/r_m]^{1/2}$, which is just the square root of the ratio of the outer and inner radii of the source distribution. In fact, from the redshifts of the x-ray Seyferts observed by the Ariel V satellite ($z \lesssim 0.1$), we have $r_m \sim \frac{1}{10} c/H_0$ in this model. It follows that the observed isotropy constraints are satisfied. More recent observations from the Einstein x-ray satellite have extended these observations to high-redshift quasars ($z \sim 2$), but this does not affect our conclusion. Nevertheless, Seyferts and quasars would not be the main contributors to the x-ray background if their evolution were such that their total x-ray emission had been significantly less in the past (a possibility which is rather less likely as a consequence of the Einstein satellite results).

An alternative view is that the bulk of the radiation may be due to the scattering of microwave radiation by 'relativistic' electrons, i.e. electrons with velocities v close to c, such that $\gamma = (1 - v^2/c^2)^{-1/2} \gg 1$. In such a scattering the energy of the electron is partly given up to the photon, so that a photon of incident frequency v becomes one of frequency of order $\gamma^2 v$. In the astrophysics literature this process is called 'inverse Compton scattering' to distinguish it from 'Compton scattering' in which high-energy photons yield up their energy to less energetic electrons, and 'Thomson scattering' of low-energy photons by non-relativistic ($\gamma \ll 1$) electrons, in which there is approximately no energy exchange. Clearly, to produce x-rays between 3×10^{17} and 10^{19} Hz from microwave photons around 3×10^{11} Hz requires Lorentz factors γ in the range $10^3 \lesssim \gamma \lesssim 3 \times 10^4$. Now it just so happens that the radio emission of the extended (double) radio sources is best explained in terms of synchrotron emission in magnetic fields of about 10^{-9} T by electrons with this range of Lorentz factors. Some x-ray emission must therefore be produced by these electrons, either within the radio lobes themselves, or by those that leak out into intergalactic space. However, it appears to be difficult to produce the whole of the observed flux with the correct spectrum in a natural way.

Rich clusters of galaxies are another class of x-ray emitters (Chapter 4); this is usually taken to indicate the existence of hot intracluster gas. The Abell catalogue provides a complete sample of rich clusters out to a distance much larger than the limit of the x-ray catalogues, and thus one can compute the contribution of the Abell clusters to the x-ray background on the assumption that, since the clusters are optically similar, their x-ray properties will also be uniform. The resultant intensity is only a small fraction of the observed background. Therefore, unless one postulates very drastic evolution of the x-ray features of clusters, so that they were either more numerous or more luminous beyond the limit of the Abell survey, rich clusters are unlikely to be responsible for a significant part of the x-ray background.

It follows from the non-thermal character of the x-ray spectrum that the x-ray background cannot have a common origin with the microwave background in a thermal equilibrium phase of the early Universe. Whatever the correct theory of their origin, it is therefore likely that the x-rays have been produced in a region much closer to us than has the microwave background, but still large compared to the length scales probed by analysis of galaxy clustering. Measurements of anisotropy in the x-ray background could then provide information on very large-scale inhomogeneities in the Universe. There is an anisotropy of about 2% in the x-ray background flux which varies with Galactic latitude and can therefore be attributed to Galactic emission. This might be due to x-rays from unidentified weak sources in the Galaxy, or from inverse Compton scattering of either microwave photons or starlight by cosmic ray electrons. Subtracting this out yields upper limits on the remaining 12 h and 24 h anisotropy of the order of 1% or less.

We can interpret this upper limit in terms of a peculiar velocity relative to the source of the x-ray background, by investigating the effect of motion on the observed intensity of a power law spectrum $i_\nu \propto \nu^{-0.4}$ (equation 3.5.1). Equation

(3.4.1) gives

$$i'_{\nu'}(\theta) \propto (1 + \beta \cos \theta)^3 \nu^{-0.4}$$

$$\propto (1 + \beta \cos \theta)^{3.4} (\nu')^{-0.4}.$$

Therefore,

$$\delta i / i = 3.4 (v/c) \cos \theta.$$

In particular, fitting Ariel V x-ray data to a $\cos \theta$ (24 h) variation about the direction determined for the velocity of the Earth by the microwave measurements of Smoot et al (1977) yields $v \lesssim 370$ km s^{-1}, which is compatible with the microwave result within the errors. Further refinement of the data should show the effect of large-scale inhomogeneities, if these exist.

3.6. Radiation and Expansion

In the big-bang picture, physical cosmology consists of the study of interacting matter and radiation expanding against their self-gravity. In this section we shall derive some of the simplest results for the way the radiation behaves in such a system under certain simplifying assumptions. Our first assumption is that we neglect the interaction of the radiation with matter; this is a good approximation at the present day when the background radiation is interacting only very weakly with matter; it turns out also to be a good representation of the early stages of the Universe, where matter and radiation were in equilibrium. It follows that our results have a wide range of validity. Secondly, we shall assume that the direct effects of gravity on the propagation of radiation are not important, an assumption which can be justified by the full theory of radiative transfer in a gravitational field (Chapter 8).

Consider then how the frequency of the radiation filling the Universe changes with expansion. To see this, note first that any individual photon we receive does not carry a marker to tell us whether it was emitted by an atom in a visible galaxy, or is part of the background radiation. Therefore the frequency of both types of photon must change in the same way. We know that, to a first approximation at least, a photon emitted by a galaxy satisfies *Hubble's law*, so that over a distance d between emission and reception, its observed wavelength changes by

$$\frac{\Delta \lambda}{\lambda} = \frac{Hd}{c} = \frac{\dot{R}}{R} t;$$

in the final expression we have substituted for the distance in terms of the light travel time, t. Now, in fact this law is valid only for short time intervals, and we shall see (§6.5) that strictly we should write a differential relation

$$\frac{d\lambda}{\lambda} = \frac{\dot{R}}{R} dt = \frac{dR}{R}$$

for the change in wavelength along an element of path dt measured by observers moving with the Hubble expansion velocity. Integrating, this gives

$$\frac{\nu_e}{\nu_0} = \frac{\lambda_0}{\lambda_e} = \frac{R(t_0)}{R(t_e)}$$

where t_0 and t_e are the times of reception and emission. This equation can be read as expressing the fact that the wavelength of a light wave expands with the Universe. For the redshift we obtain the fundamental result

$$1 + z = \frac{R(t_0)}{R(t_e)}.$$

From our initial remark, this law must be equally valid for the background radiation. Note, however, that the result has been suggested by the locally observed form of the Hubble law, not proved.

We can now show that a black-body spectrum is maintained under expansion. The Planck function tells us how many photons, $n_\nu \, d\nu$, per unit volume of space occupy a given frequency interval $d\nu$ in thermal equilibrium at temperature T, viz.

$$n_\nu \, d\nu = \frac{8\pi}{c^3} \frac{\nu^2 \, d\nu}{\exp(h\nu/kT) - 1}.$$

As such a radiation field expands, the change in frequency relabels the modes, but, under the assumption that gravity has no direct effect, no new photons are created and none destroyed. Therefore, at a later time t the number of photons in the mode labelled by $\nu(t) = R(t_0)/R(t)\nu_0$ in a given region of space equals that originally in the mode labelled by ν_0. Since a given region of space increases in volume by a factor R^3/R_0^3, where, for simplicity, we write $R = R(t)$ and $R_0 = R(t_0)$, we have

$$\frac{8\pi}{c^3} \frac{\nu_0^2 \, d\nu_0}{\exp(h\nu_0/kT_0) - 1} = \frac{8\pi}{c^3} \frac{R^3}{R_0^3} \frac{\nu^2 \, d\nu}{\exp(h\nu/kT) - 1}$$

$$= \frac{8\pi}{c^3} \frac{R^3}{R_0^3} \left\{ \frac{R_0^3}{R^3} \frac{\nu_0^2 \, d\nu_0}{\exp(h\nu_0 R_0/RkT) - 1} \right\}$$

$$= \frac{8\pi}{c^3} \frac{\nu_0^2 \, d\nu_0}{\exp(h\nu_0 R_0/RkT) - 1}.$$

Therefore

$$R_0/RT = 1/T_0$$

or

$$T(t) \propto 1/R(t). \tag{3.6.1}$$

We see that the Planck spectrum is unchanged in form, but the radiation cools according to (3.6.1).

Notice that the actual identity of the photons in the region of space considered in this argument will have changed between the two times t_0 and t: it is only the numerical values of the occupation numbers of the modes that do not change. It is usual to avoid unnecessary worry on this score by confining attention in these arguments to a small portion of the Universe marked off by a hypothetical box with perfectly reflecting walls which expands with the Universe. The situation inside the box is exactly as it would have been in the absence of the box, since the assumed homogeneity of the system ensures that every photon that should have left the box is reflected into one that should have entered. Now, however, the actual identity of the photons is maintained in time.

As a simple application of this result consider how the energy density of radiation changes with time. Since we are dealing with black-body radiation, the energy density is $u = aT^4$ and so, from (3.6.1),

$$u \propto 1/R^4.$$

Note that this result is consistent with the maintenance of a Planck spectrum and the relabelling of the modes, as indeed it must be, for we have

$$u_\nu \, d\nu \propto \frac{\nu^3 \, d\nu}{\exp(h\nu/kT) - 1} \propto \frac{1}{R^4} \frac{\nu_0^3 \, d\nu_0}{\exp(h\nu_0/kT_0) - 1}.$$

Therefore the energy density in each band $d\nu$ is proportional to R^{-4}, and so is the total over all bands.

Since the volume of the expanding box is proportional to R^3, this means that the total energy in the box decreases on expansion. This energy goes into work done by the pressure of radiation against the gravitational field during expansion. For the work done in expanding by dR is

$$p_r \, dV = \tfrac{1}{3} u \, d(R^3) \left(\frac{V_0}{R_0^3} \right) = R^2 u \, dR \left(\frac{V_0}{R_0^3} \right), \tag{3.6.2}$$

where $V = (R/R_0)^3 V_0$ is the volume at time t in terms of that at t_0, and $p_r = \tfrac{1}{3} u$ is the radiation pressure. The change in internal energy is

$$d(uV) = d \left(\frac{u_0 V_0 R_0}{R} \right) = - \, dR \, \frac{u_0 V_0 R_0}{R^2} = - uR^2 \, dR \left(\frac{V_0}{R_0^3} \right)$$

which balances equation (3.6.2).

This means that the heat input into the system must be zero, by application of the second law of thermodynamics to the radiation in the box; since the box is an isolated system this is to be expected. We can confirm it by computing the entropy of the radiation. Let s be the entropy per unit volume. Then the second law of thermodynamics gives

$$d(uV) = T \, d(sV) - p_r \, dV$$

for a volume V. Using $p_r = \frac{1}{3}u$, this is

$$(du - T\,ds) = (Ts - \frac{4}{3}u)\,dV/V.$$

Since u and s are independent of V, both sides of this equation must vanish. This gives two equations which are each equivalent to

$$s = \frac{4}{3}aT^3.$$

The total entropy in volume V is $sV \propto T^3V$, and from equation (3.6.1), this is constant as the Universe expands. This is the required result.

A Universe filled with radiation expands as an adiabatic gas of photons. Since $p \propto u \propto R^{-4}$ and $V \propto R^3$, we have $pV^{4/3} = $ constant, so the gas has adiabatic index 4/3, which is just the usual result for a photon gas. The thermodynamics of the expanding universal radiation field is therefore precisely that of its 'laboratory' counterpart.

4. *The Quantity of Matter*

4.1. The Masses of Galaxies

The masses of astronomical objects can be determined directly only through the gravitational forces they produce. The forces induce accelerations, and although these cannot themselves be measured, they give rise to velocities which can be detected by Doppler shifts. If certain assumptions are made to relate the observed velocities to the actual accelerations, masses can be estimated.

The simplest example is provided by the 2% or so of galaxies that occur in binary systems. In the case of binary galaxies, in contrast to the situation for binary stars, only the instantaneous velocities can be measured, and so no information can be obtained on the inclination of the orbital plane. Nevertheless, by averaging over a large number of examples, and assuming random orientations, an average mass can be calculated. Binary systems contain all morphological types of galaxies, in both similar and dissimilar pairs, with a number distribution among the types resembling that of non-cluster galaxies in general. The results (Page 1975) indicate that the average mass of ellipticals is some 30 times the average spiral mass, with the latter of order $2 \times 10^{10} M_\odot$.

This approach may be extended to bound systems containing many galaxies and the average mass of the members may be obtained by using the *virial theorem* which is derived in §4.2. This effectively relates the observed spatial distribution of the galaxies to their relative velocities. The method is considered further in §4.4 as it applies to rich clusters of galaxies, and in relation to small groups in §4.5.

In the case of non-rotating elliptical (EO) galaxies, the velocity dispersion of stars can be used to provide a measure of the random velocities in the centre of the galaxy, since these are generated in the gravitational potential of the galaxy. The mass of the galaxy can then be obtained from the virial theorem, as outlined in §4.3.

In rotating galaxies the rotation of one component may be measured (the stars or neutral hydrogen, for example), and the results fitted to a model mass distribution. For illustration, assume the whole mass of the galaxy, M, to be concentrated at a central mass point. At radius r, the rotational velocity, v, is given by

$$\frac{v^2}{r} = \frac{GM}{r^2}.$$

Measurement of v near the edge of the galaxy gives a crude estimate of M. In practice, more realistic spheroidal distributions or disc models are used, and the mass distribution computed from complete rotation curves, rather than simply from the velocity at a single point. This method has been used in conjunction with rotation curves derived from stellar absorption lines, emission lines from regions of ionised hydrogen, and from 21 cm radio emission from neutral hydrogen.

From the photographic magnitude of a galaxy the mass to light ratio, M/L, can be determined. This ranges from less than the solar value, $M_\odot/L_\odot$, for some Sc galaxies at one extreme, to perhaps more than $70\,M_\odot/L_\odot$ for some ellipticals. In general, for spirals one finds $1 < (M/L)/(M_\odot/L_\odot) < 15$, and for ellipticals, $10 < (M/L)/(M_\odot/L_\odot) < 70$. Distance estimates, and hence the values of mass and luminosity, depend on the Hubble constant. If $H_0 = 100\,h$ km s^{-1} Mpc then, for the purpose of order of magnitude estimates at least, an average mass to light ratio can be taken to be $20\,h\,M_\odot/L_\odot$.

4.2. The Virial Theorem

The *virial theorem* is a conservation law for systems of interacting particles which have achieved a state of equilibrium. It is therefore of less general applicability than the conservation of energy and momentum which are valid whether or not a system is in equilibrium. If T is the total kinetic energy and V the total potential energy of the particles, the virial theorem states that

$$2T + V = 0. \qquad (4.2.1)$$

We derive this theorem for the case of N point particles acting gravitationally upon each other. Let the ith particle have mass m_i and position vector $\mathbf{r}_i$ relative to an arbitrarily chosen origin at rest. Then the equation of motion for the jth particle is

$$m_j \frac{d^2\mathbf{r}_j}{dt^2} = \sum_{i \neq j} \frac{Gm_im_j}{|\mathbf{r}_i - \mathbf{r}_j|^3}(\mathbf{r}_i - \mathbf{r}_j), \qquad (4.2.2)$$

since the right-hand side is the vector sum of the gravitational force on the jth particle due to all the other particles. Taking the scalar product with $\mathbf{r}_j$, and summing over j, gives, on the left-hand side,

$$\sum_j m_j \frac{d^2\mathbf{r}_j}{dt^2} \cdot \mathbf{r}_j = \sum_j m_j \frac{d}{dt}\left(\frac{d\mathbf{r}_j}{dt} \cdot \mathbf{r}_j\right) - \sum_j m_j \left(\frac{d\mathbf{r}_j}{dt} \cdot \frac{d\mathbf{r}_j}{dt}\right)$$

$$= \frac{d^2}{dt^2} \sum_j \tfrac{1}{2}m_j(\mathbf{r}_j \cdot \mathbf{r}_j) - \sum_j m_j \left(\frac{d\mathbf{r}_j}{dt} \cdot \frac{d\mathbf{r}_j}{dt}\right) \qquad (4.2.3)$$

The first term on the right is the second derivative of the moment of inertia, which is zero if the system is in equilibrium. The second term is just $-2T$. On the right-hand side of (4.2.2) we take the scalar product with $r_j = \frac{1}{2}(r_i + r_j) - \frac{1}{2}(r_i - r_j)$ to get

$$\frac{1}{2}\sum_j \sum_{i \neq j} \frac{Gm_i m_j}{|r_i - r_j|^3}(r_i - r_j)\cdot(r_i + r_j) - \frac{1}{2}\sum_j \sum_{i \neq j} \frac{Gm_i m_j}{|r_i - r_j|}. \qquad (4.2.4)$$

The first term is zero by symmetry, since each term in the sum occurs twice with opposite sign; for example $i = 1$, $j = 2$ and $i = 2$, $j = 1$ give cancelling terms. The second term in (4.2.4) is the gravitational potential energy of the system, V. Therefore, the surviving terms in (4.2.3) and (4.2.4) yield the result $2T + V = 0$, as required.

4.3. Application to Galaxies

As stated in §4.1, the virial theorem can be used to provide an estimate of the mass of a non-rotating galaxy. The kinetic energy, T, can be determined in terms of the velocity dispersion, $\langle v^2 \rangle$, of the central stars as measured from the broadening of spectral lines. Since only velocities along the line of sight contribute, two out of the three directions make no contribution to the measured value, so the true velocity dispersion is three times the measured one. Thus we write $2T = 3M\langle v^2 \rangle$. The potential energy can be related to the distribution of surface brightness across the galaxy, under the assumption that the mass to light ratio is constant, since the brightness distribution then gives a direct measure of the mass distribution. It can be shown that $V \approx -\frac{1}{3}GM^2/R_{1/2}$, where $R_{1/2}$ is the radius of a circle which encloses half of the luminosity of the galaxy. Clearly the virial theorem (4.2.1) yields

$$M \sim \frac{9R_{1/2}\langle v^2 \rangle}{G}.$$

Several elliptical galaxies have had their mass to light ratios measured in this way. Less massive systems ($\lesssim 10^{11} M_\odot$) typically have $M/L \sim 5 M_\odot/L_\odot$, comparable with spirals, while some of the more massive galaxies ($\gtrsim 10^{12} M_\odot$) give $M/L \gtrsim 50 M_\odot/L_\odot$. It is possible to take the rotation of galaxies into account by adding an extra rotational kinetic energy term in the virial theorem. In practice, however, there is a problem in obtaining the required rotation rate from the observed ellipticity of a galaxy; indeed, it is quite probable that the ellipticity of elliptical galaxies is not due to rotation at all.

4.4. Rich Clusters of Galaxies

From the redshifts of the individual member galaxies of a cluster relative to their mean Hubble redshift, typical random velocities of cluster members can be deduced. It is then found that the time required for a typical galaxy to cross the cluster is much less than the age of the Universe; thus, if the cluster were not a bound system there would have been ample time for the galaxies to disperse.

In addition, to a first approximation at least, the regular spherical distribution of galaxies throughout a rich cluster, such as Coma, resembles the distribution of atoms in a finite volume of an isothermal sphere of self-gravitating gas. Consequently, we may assume that such systems are in equilibrium and apply the virial theorem in order to determine the mean mass of the constituent galaxies.

If the ith galaxy has mass m_i and velocity v_i, the kinetic energy is given by

$$\tfrac{1}{2}\sum m_i v_i^2 = \tfrac{3}{2}N\bar{m}\langle v^2\rangle, \tag{4.4.1}$$

where N is the total number of galaxies of average mass $\bar{m}$, and $\langle v^2\rangle$ is the mean square peculiar velocity, or velocity dispersion, relative to the recessional velocity of the cluster as a whole. The factor of 3 appears again since $\langle v^2\rangle$ is the measured velocity along the line of sight, transverse velocities contributing nothing to the redshift, whereas the kinetic energy depends on the velocity in space.

For a spherical cluster, the surface density of galaxies $\sigma(R)$ in a ring of radius R projected on the sky is related to the volume number density $n(r)$ by

$$\sigma(R) = 2 \int_R^a n(r) \, \frac{r \, dr}{(r^2 - R^2)^{1/2}} \, .$$

This is a form of Abel's integral equation, and can be inverted to give the number density of galaxies in terms of observable quantities:

$$n(r) = \tfrac{1}{2} \int_r^a \frac{d\sigma}{dR} \, \frac{R \, dR}{(r^2 - R^2)^{1/2}} \, .$$

The potential energy is then obtained from

$$-\frac{V}{\bar{m}^2} = G \int_0^R \frac{n(r)}{r} \, dn(r) = G \int_0^R \left\{ \int_0^r \frac{n(r')4\pi r'^2 \, dr'}{r} \right\} 4\pi r^2 n(r) \, dr. \tag{4.4.2}$$

The virial theorem (4.2.1), (4.4.1) and (4.4.2), together with the observed values give, for Coma,

$$N\bar{m} = 1.5 \times 10^{15} h^{-1} M_\odot,$$

or a mean mass to light ratio $\bar{m}/L \sim 400h \, M_\odot/L_\odot$.

Even allowing for the fact that the Coma cluster contains a preponderance of ellipticals, which have the largest mass to light ratio (§4.1), this result differs significantly from our previous conclusion. The problem was first noted by Zwicky and by Smith for the Virgo cluster, and has been subsequently discussed by many authors for other rich clusters. It is called the *missing mass problem* since it can be interpreted as implying a deficiency of mass (as determined by multiplying the supposedly known mean masses of individual galaxies by the number of galaxies in the cluster) with respect to the mass determined from the dynamics of the systems through the virial theorem.

Note that the missing mass is larger, by perhaps even a factor of 10 according to the above analysis, than the mass in galaxies, and so does not represent a small correction! The observed velocity dispersion across the cluster suggests that the gravitational potential varies as would be expected for a mass distribution similar to the observed distribution of galaxies. The missing mass must therefore be concentrated towards the centre, as are the galaxies, and this constrains the possibilities for the missing matter. For example, it would be expected from this argument that any intracluster gas has a temperature corresponding to the velocity dispersion found for the galaxies, $T = \frac{1}{2}m_H \langle v^2 \rangle / k \sim 6 \times 10^7 \,\mathrm{K}$, in order that it be held up in the same way as the galaxies in the gravitational potential well of the cluster. At this temperature the central region should be a powerful x-ray source. It should cool in much less than the age of the Universe (and so, presumably, in much less than the age of the cluster) unless some reheating processes can be postulated. In fact, x-ray observations indicate that hot intracluster gas provides no more than 1% of the missing mass.

Alternative forms for intracluster material have been suggested which avoid these difficulties. For example, matter in the form of black holes, provided these are small enough not to disrupt the galaxies by tidal effects, could provide the missing mass. Alternatively, lumps of solid hydrogen in sufficient numbers would appear to be stable and would have avoided detection. In both cases, there is then the problem of explaining how the missing mass could have arrived in its postulated form. The existence of massive halos to the galaxies could resolve the problem, but direct, independent evidence for such halos in cluster members is lacking. This possibility is discussed again briefly in §4.5. Explanations of the discrepancy other than in terms of missing matter are possible. One possibility is that observed clusters may not, in fact, be single dynamical entities. The Virgo cluster may be two clusters; and the Hercules cluster may include overlapping smaller clusters with the result that the velocity dispersion is spuriously high.

At present, the missing mass problem is unresolved. The consensus view is that the matter is not missing, but is present in some form. The problem is of considerable cosmological importance since the upper limits for the missing mass may be sufficient to alter qualititatively the evolution of the Universe; if the maximum allowable mass were indeed present, then, according to our current understanding (Chapter 6), the Universe would be doomed to a future recontraction.

4.5. Small Groups of Galaxies

A direct application of the virial theorem to small clusters of galaxies, such as the Local Group, also gives discrepant results for the mass of the system. The Local Group is dominated by the Andromeda galaxy and our Galaxy, which are now approaching each other with a velocity of $90 \,\mathrm{km\ s^{-1}}$. A simple interpretation, neglecting the other galaxies in the group, is that the gravitational force between these two galaxies is sufficient to halt and reverse the initial Hubble recession at the time of formation. This can only be the case if the mass is sufficient, as can be seen from a simplified estimate. The separation, r, of the two galaxies at any time satisfies

$$\frac{d^2r}{dt^2} = -\frac{GM}{r^2},$$
(4.5.1)

where M is the mass of the system. Estimating d^2r/dt^2 as R/t^2, where R is the present separation (660 kpc) and t is the time since formation (which can be taken to be of order H_0^{-1} or 10^{10} years), we obtain

$$M \sim R^3/Gt^2 \sim 1.3 \times 10^{12} M_\odot.$$

This is at least an order of magnitude greater than the mass of spiral galaxies determined from rotation curves, and the discrepancy is not due to the crudeness of this estimate.

If this indicates the existence of missing matter, one possibility is the existence of massive, spherically symmetric halos to spiral galaxies. A rotation curve analysis to radius r gives no information on a spherical distribution of matter outside r, and direct contradiction with observation can be avoided by having additional matter in some low-luminosity form such as late-type dwarf stars. Some evidence for the existence of such halos has been claimed in the orbits of globular clusters. In addition, there is an independent theoretical reason for postulating a massive halo: the spiral disc is apparently unstable, and could be stabilised by the presence of a halo. On the other hand, the rotation curve results seem to agree reasonably well with the average masses determined from binary systems, which yield, of course, the total masses. However, the analysis of binary systems is not without its critics.

Alternatively, it is possible to argue that the observations of small groups support the view that the Universe has a low density, with essentially no hidden mass not in galaxies. The argument proceeds by noting that in the derivation of the virial theorem no account has been taken of the Hubble expansion. Thus, it is claimed, the standard form of the virial theorem is inappropriate since it applies to a closed, isolated system, and not to loosely bound systems which can be affected by the overall expansion of the Universe. In this case, the usual virial theorem is to be replaced by a 'statistical' virial theorem, relating to the departures from a smooth, uniformly expanding density of matter.

To obtain some insight into how this idea works, consider the equations of motion of a galaxy with peculiar velocity v at a distance r from our Galaxy. For simplicity assume v is radial. Then by definition, and Hubble's law, we have

$$\frac{dr}{dt} = v + Hr.$$
(4.5.2)

The changes in velocity are to be measured relative to the Hubble flow, that is, the imaginary trajectories of particles expanding uniformly in a smoothed-out distribution of matter. This manifests itself in an additional term in the acceleration equation, i.e.

$$\frac{dv}{dt} + Hv = -\frac{GM}{r^2} \qquad (4.5.3)$$

[c.f. equation (4.5.1); for a derivation of (4.5.3) see Layzer 1975]. These equations of motion no longer determine the mass of the system M. Instead, for a given M, they determine, in principle, the time dependence of the Hubble parameter, H, which would lead to any desired trajectory for the galaxy under discussion. We shall see in Chapter 6 that this is equivalent to determining the mean density of the Universe.

Multiplying (4.5.3) by v and using (4.5.2) gives

$$\frac{d}{dt}\left(\tfrac{1}{2}v^2 - \frac{GM}{r}\right) + H\left(v^2 - \frac{GM}{r}\right) = 0. \qquad (4.5.4)$$

Therefore, *if* the energy of the two-particle system is conserved, so $\tfrac{1}{2}v^2 - GM/r$ = constant, the virial theorem will be valid for the two particles, $v^2 - GM/r = 0$. *But*, for a non-isolated system, energy is not necessarily conserved and the virial theorem is not, in fact, satisfied.

Using $H = \dot{R}/R$, equation (4.5.4) can be rearranged to give

$$\frac{1}{R}\frac{d}{dt}\left(\tfrac{1}{2}Rv^2 - \frac{GMR}{r}\right) = -\tfrac{1}{2}Hv^2 < 0.$$

The right-hand side is less than zero, so $R(\tfrac{1}{2}v^2 - GM/r)$ is a decreasing function; since R is increasing, this means $(\tfrac{1}{2}v^2 - GM/r)$ is decreasing. Returning now to equation (4.5.4) we deduce that

$$v^2 - GM/r > 0,$$

or

$$2T + V = T_0 > 0.$$

It follows that

$$M \propto -\frac{V}{M} = \frac{2(T-T_0)}{M},$$

and hence that the estimate of the mass of the system is decreased. A more rigorous analysis shows that the discrepancy can be removed for small groups by a suitable choice of $H(t)$ (Fall 1976). This does not, of course, prove that this is the way Nature chooses to keep itself consistent.

4.6. The Mass Density of the Universe

The amount of matter in the world is a quantity of some considerable importance in cosmology since, as we shall see in Chapter 6, it determines, through the effect of gravity, the way in which the Universe evolves. Indeed, the mass density of the Universe, ρ, is often expressed in units of a *critical density*, $\rho_c = 1.9 \times 10^{-26} h^2$ kg m^{-3}, and a *density parameter*, $\Omega = \rho/\rho_c$. In the simplest models (§6.7), a universe with less than the critical density expands for ever, while if $\Omega > 1$ the universe is destined to recollapse.

From our determination of typical values for the masses of galaxies we can find the amount of matter in visible galaxies: this must provide a lower limit to the average density of the Universe. A further contribution could come from matter with the same light output per unit mass as normal galaxies, but organised into dwarf galaxies too faint to be seen even if relatively nearby. We can estimate a limit to the amount of matter in this form by requiring the integrated emission to be less than the observed optical background. In addition, there may be large amounts of relatively dark material, provided this is arranged in suitably cunning ways. Examples in this category are the still elusive intergalactic medium or the exotica of small black holes, or judiciously proportioned snowflakes of solid hydrogen. We may expect that the density of the intergalactic medium will be determined in due course by sufficiently accurate observations. On the other hand, it is unlikely that any direct limits can be set on the more contrived forms, and for this we must rely on indirect evidence, such as the abundance of deuterium, or the clustering of galaxies, or the deceleration of the expansion of the Universe, within the context of specific theories, as will be described later.

The mass density in galaxies cannot be found directly by multiplying together an average galactic mass and a mean number density obtained by counting, since we cannot be certain of counting *all* galaxies out to a given distance; only the brighter galaxies will be counted at the further distances. To compensate for this we should need to use the luminosity function for galaxies, which is very uncertain for faint galaxies. It is, in any case, easier to use the information on luminosities more directly. Thus, we obtain limits to the mass density from observations of the total light output per unit volume. For, if galaxies radiate into space an energy ℓ_g per second per unit volume in some waveband, and if M/L is the average mass to light ratio of galaxies in that waveband, then

$$\rho_g = (M/L)\, \ell_g \qquad (4.6.1)$$

gives the average density of matter in galaxies at the present epoch.

In order to obtain a lower limit for ρ_g we compute ℓ_g on the basis of a contribution from galaxies only of the types we can actually see in our local environment. In practice, the shape of the galaxy luminosity function is usually obtained from the Coma cluster and normalised by fitting to observed counts of galaxies to given limiting magnitudes (§1.2). If $(d\Phi(M)/dM)\, dM$ is the number of galaxies per unit volume with absolute magnitudes between M and $M + dM$, then

$$\ell_g = \int_{-\infty}^{\infty} \frac{d\Phi}{dM}\, L(M)\, dM,$$

where $L(M) = 10^{0.4(4.79-M)} L_\odot$ is the luminosity of a galaxy of magnitude M. In conjunction with equation (4.6.1), and a representative value of the mass to light ratio of $M/L = 20\, h\, M_\odot/L_\odot$ (§4.1) this yields a lower limit

$$\rho_g \gtrsim 4 \times 10^{-28} h^2 \text{ kg m}^{-3}.$$

To obtain now an upper limit for the mass density in galaxies, we attribute all the brightness of the night sky that cannot definitely be accounted for otherwise to light from galaxies with a standard mass to light ratio. In principle, it is necessary to take into account effects due to the expansion of the Universe and also to the evolution of the stellar content of galaxies in order to compute ℓ_g from the observed sky brightness. However, the main cosmological effect is that energy received in the optical band must have been emitted at increasingly shorter wavelengths in the ultraviolet as we look to higher redshifts. Since normal galaxies do not emit significantly in the ultraviolet, we can neglect the contribution from those beyond a relatively modest redshift, say $z \sim 1$. In these circumstances, we can obtain a reasonable order of magnitude estimate by treating the Universe as a static sphere of radius c/H_0, i.e. such that the light travel time across the sphere equals the present age. One can see this by noting that the cosmological effects introduce factors of $R(t_0)/R(t_e) = (1+z)$ to some relatively modest power. For example, evolution in the number density of galaxies, $n(t)$, gives $n(t) \propto R(t)^{-3} \propto (1+z)^{-3}$, since a fixed number of galaxies occupy a volume which increases as R^3. These effects are therefore small for small z.

The energy density in this spherical static model of the Universe is ℓ_g/H_0, so the sky brightness in the relevant band is given by

$$i = \frac{c}{4\pi} \cdot \frac{\ell_g}{H_0} = \frac{c}{H_0} \frac{\ell_g}{4\pi}.$$

The final expression simply represents i as the integral of the emission per unit volume per unit solid angle, $j = \ell_g/4\pi$ (W m^{-3} sr^{-1}), along the line of sight.

Observations of the brightness of the night sky (§3.1) yield an upper limit on the extragalactic component of about $\nu i_\nu \approx 3 \times 10^{-8}$ W m^{-2} sr^{-1}. Taking $i \sim \nu i_\nu$, this gives $\ell_g \lesssim 4 \times 10^{-59} L_\odot$ W m^{-3} Mpc^{-3}; and for a mass to light ratio $M/L = 20\, h$ $M_\odot/L_\odot$ as before, we obtain

$$\rho_g \lesssim 10^{-27} h^2 \text{ kg m}^{-3}.$$

4.7. The Intergalactic Medium

The positive detection of gas between galaxies in clusters has already been mentioned in §4.4. This suggests that the possibility of gas between clusters should be investigated, especially as it is difficult to imagine that the process of galaxy formation could have been so efficient that it did not leave some gas outside clusters. In fact, no detection of such gas has been made, so the most we can do at present is to use the observed null results of various tests to set upper limits on the intergalactic gas density. We do this on the assumption of a uniform medium; this is not realistic since it is improbable that the gas should be excluded from the clustering tendency of galaxies. Nevertheless, since no gas has so far been detected, it is pointless to investigate arbitrarily myriad possibilities which do not exist (even if by accident one might happen to include the one that does).

The various tests can be divided into two types: tests for emission of radiation from the gas, or for absorption and related phenomena such as dispersion. Furthermore, the gas may be neutral or ionised, and the ionisation may be achieved either thermally or by photoionisation. Most of the tests look for the presence of hydrogen, since this is presumably the most abundant element. We shall investigate order of magnitude limits for the mean mass density of the intergalactic medium provided by some of the more important of these tests.

4.8. Emission by Neutral Hydrogen

The interaction between the spins of the electron and of the proton produces a splitting in the ground state of the hydrogen atom of energy $\Delta E = 6\ \mu eV$, corresponding to radiation at 1428 MHz or 21 cm. If the hydrogen atoms were in equilibrium with thermal radiation at temperature T, the upper and lower levels would be populated with number densities n_2, n_1, respectively, given by the Boltzmann formula

$$\frac{n_2}{n_1} = \frac{g_2}{g_1} \exp\left(-E/kT\right). \tag{4.8.1}$$

The ratio of statistical weights, g_2/g_1, is here equal to 3, since the upper state is a triplet and the lower one a singlet. The microwave background provides the required thermal radiation field at 2.7 K. In decaying from the excited state any neutral hydrogen produces emission at 21 cm. If the matter and radiation were indeed in equilibrium, however, this emission would be balanced by absorption, and the presence of the matter would not be detectable in this way. On the other hand, thermal equilibrium can only be established for large optical depths in the gas, and hence for large amounts of neutral hydrogen. We can therefore consider the opposite case when the gas is optically thin and emission is not followed by absorption. The lack of observed emission can then be used to limit the density of emitting material and the consistency of the discussion confirmed by estimating the corresponding optical depth. In fact, we know from observations that the medium is not optically thick to 21 cm emission out to the radio galaxy Cygnus A, and we shall use this information in §4.9 to provide another limit on the intergalactic hydrogen density.

If thermal equilibrium does not exist, then the relative populations of the levels are not given by the Boltzmann formula (4.8.1). In this situation we *define* a temperature T_s, usually referred to as a 'spin temperature', through a formally similar equation

$$\frac{n_2}{n_1} = \frac{g_2}{g_1} \exp\left(-E/kT_s\right). \tag{4.8.2}$$

We must have $T_s \geqslant 2.7$ K since the background photons at 21 cm are available to excite the atoms. The only alternative, that this radiation had not had sufficient time to achieve significant excitation in the available lifetime of the Universe, can be shown to be false. The case $T_s \approx 2.7$ K would imply approximate thermal equilibrium, which we are not considering. Thus we require $T_s \gg 2.7$ K, and there must be a source of excitation in addition to the microwave background. This could be, for example, collisions between atoms. In fact, the principal mechanism is more complicated, but, since we do not need to know here how a larger T_s is achieved, we shall reserve discussion till §4.9. From these assumptions it follows that $T_s \gg \Delta E/k = 0.07$ K. This means we can take $n_2/n_1 \approx g_2/g_1 = 3$ in place of equation (4.8.2).

The decay of an atom in its excited state takes place after an average lifetime of $A^{-1} = 3.5 \times 10^{14}$ s. At a time when the total number density of hydrogen atoms is n, the transition rate to the ground state per unit volume is $\frac{3}{4} An$, since $\frac{3}{4}$ of the atoms are in the excited state. The total emission coefficient (the energy emitted per unit volume per second per unit solid angle at all frequencies) is

$$j = (1/4\pi)\tfrac{3}{4} Anh\nu_\varrho \, \text{W m}^{-3} \text{sr}^{-1}, \tag{4.8.3}$$

and the energy is emitted in a narrow line around $\nu_\varrho \sim 1428$ MHz measured at the point of emission. The emission coefficient is a function of time through n. At some time, t, in the past the intergalactic medium would have been denser than it is now, at t_0, by a factor $[R(t_0)/R(t)]^3$, and the emission coefficient would have been correspondingly greater. For ease of notation we shall again write $R_0 = R(t_0)$ and $R = R(t)$.

The received radiation is, of course, redshifted by the expansion of the Universe. This has two effects: first, according to §3.6, the received frequency is given by

$$\nu = \frac{R}{R_0} \nu_\varrho, \tag{4.8.4}$$

for radiation emitted at time t, and the photons are received in a line with a width increased by the same factor. Secondly, the number of photons per unit volume of space at the receiver at this frequency is less than that in the emission region at time t by a factor $(R/R_0)^3$. This follows because the number of these particular photons is unchanged in time, but the volume through which they are spread increases by $(R_0/R)^3$; the energy density of the radiation is therefore diminished by $(R/R_0)^4$ between emission and reception, and the energy density per unit band width by $(R/R_0)^3$. For an isotropic distribution of radiation the intensity is simply related to the energy density by a factor $c/4\pi$, so the intensity suffers the same diminution with expansion.

We assume that re-absorption in the intergalactic gas can be neglected, i.e. that the gas is optically thin. This can be shown to be consistent with the derived limit. The received intensity $i_\nu \, d\nu$ in the element of bandwidth $d\nu$ can now be calculated. Since $i_\nu \, d\nu$ is the power per unit area of sky, it is obtained from the power per unit volume, j, by integration through the depth of sky making a contribution in the given $d\nu$, with due allowance for the expansion factors. In fact, only an element of the medium along the line of sight will contribute to $i_\nu \, d\nu$, since only an element of the path will have the correct redshift to put the given emission frequency ν_ϱ into the range $(\nu, \nu + d\nu)$. From equation (4.8.4) we have

$$d\nu = \frac{\nu_\varrho}{R_0} \dot{R} \, dt,$$

so the appropriate element of path, $ds = c \, dt$, is given by

$$ds = \frac{c \, d\nu}{\nu H},$$

where $H = \dot{R}/R$ is, as usual, the Hubble parameter at time t. Therefore

$$i_\nu \, d\nu = \left(\frac{R}{R_0}\right)^3 j \, ds = \left(\frac{R}{R_0}\right)^3 \frac{jc}{\nu H} \, d\nu \qquad \nu < \nu_\varrho$$

$$= 0. \qquad\qquad \nu > \nu_\varrho$$

We conclude that at $\nu = \nu_\varrho = 1428$ MHz, the radio brightness of the sky should change by

$$i_{\nu_\varrho} = \frac{1}{4\pi} \cdot \tfrac{3}{4} A \, \frac{n_0 h c}{H_0},$$

where n_0 and H_0 are present values. Observations yield $i_{\nu_\varrho} < 5.4 \times 10^{-11}$ W m^{-2} Hz^{-1} sr^{-1}, from which we compute a limit to the neutral hydrogen density

$$\rho_{HI} \lesssim 8 \times 10^{-28} h \text{ kg m}^{-3}.$$

This limit is comparable to the mass in galaxies, but we shall see below that more stringent limits can be obtained.

4.9. Absorption by Neutral Hydrogen

The emission from neutral hydrogen may be detected, at least in principle, from dark areas of the radio sky. In contrast, by looking at radiation from extragalactic radio sources one expects to observe the effect of absorption. An upper limit for the optical depth to the double radio galaxy Cygnus A has been determined to be $\tau < 5 \times 10^{-4}$ at 21 cm from the absence of a discontinuity in its radio spectrum attributable to intergalactic absorption. Our aim is to relate this to an upper limit for the density of an intergalactic medium of neutral hydrogen.

Since we already have the emission coefficient (4.8.3), the simplest way to obtain the opacity, κ_ν, is to use the Kirchhoff relation, $\kappa_\nu = j_\nu/B_\nu(T_s)$, where $B_\nu(T_s)$ is the Planck function at the spin temperature, and j_ν is the emission coefficient at frequency ν. The average opacity over the line of width $\Delta\nu$ at $\nu_\varrho = 1428$ Hz is

$$\bar{\kappa} = \frac{1}{\Delta\nu} \int_{\nu_\varrho - \Delta\nu/2}^{\nu_\varrho + \Delta\nu/2} \kappa_\nu \, \mathrm{d}\nu$$

$$= \frac{1}{\Delta\nu} \int_{\nu_\varrho - \Delta\nu/2}^{\nu_\varrho + \Delta\nu/2} j_\nu/B_\nu \, \mathrm{d}\nu.$$

Since B_ν varies only slowly through the line, while j_ν varies from zero at the edges to a maximum in the centre, we can take B_ν out of the integrand to get

$$\bar{\kappa} = \frac{1}{\Delta\nu B_{\nu_\varrho}} \int j_\nu \, \mathrm{d}\nu = \frac{j}{\Delta\nu B_{\nu_\varrho}}, \qquad (4.9.1)$$

with j now given by (4.8.3).

Again we take $T_s \gg \Delta E/k$, since T_s is at least 2.7 K, this means we can use the Rayleigh–Jeans approximation to the Planck function, $B_\nu(T_s) \approx (2k\nu^2/c^2)\,T_s$, to obtain

$$\bar{\kappa}\Delta\nu = \frac{3}{32\pi} nAh\nu_\varrho \frac{c^2}{kT_s\nu_\varrho^2}.$$

Absorption at a given observed frequency occurs at a time in the past along the line of sight such that the given frequency is in the line at 1428 MHz. This is possible for all observed frequencies less than 1428 MHz, but not for higher frequencies, so we expect a discontinuity in the absorption. The optical depth at $\nu < \nu_\varrho$ is therefore $\bar{\kappa}c\Delta t$, where, as in §4.8, $c\Delta t = c\Delta\nu/H\nu$; hence

$$\tau_\nu = (\bar{\kappa}\Delta\nu)\,c/H\nu \qquad \nu < \nu_\varrho.$$

The step at 1428 Mhz is obtained by putting $H = H_0$, and $\nu = \nu_\varrho$, i.e.

$$\tau_{\nu_\varrho} = \frac{3}{32\pi} \frac{Ahc^3}{kT_s} \frac{n_0}{\nu_\varrho^3} \frac{1}{H_0},$$

which is the required result. From the Cygnus A data we obtain

$$\rho_{HI} \lesssim 5 \times 10^{-30}\, T_s \text{ kg m}^{-3}.$$

To obtain a realistic upper limit we need to estimate T_s. This is determined by the amount to which processes other than absorption from the microwave background contribute to the population of the excited level. The major contribution turns out to come from the absorption of ultraviolet radiation at the wavelength of the Lyman-α line (1215 Å) by a hydrogen atom in the lower level of the ground state. This excites the atom to the $2P_{3/2}$ level, and is followed by decay to the upper (triplet) level of the ground state. In equilibrium, this would be balanced by the reverse process in which an electron in the upper level is excited by a Lyman-α photon and decays subsequently to the lower (singlet) ground state. But the atoms are also in contact with radiation at 2.7 K, so they arrange themselves with T_s somewhere between 2.7 K and a temperature characterising the ultraviolet continuum. The precise value corresponds to that at which the rates of the different processes are in equilibrium and can be calculated accordingly. Measurements of the strength of the ultraviolet background then yield $T_s \lesssim 60$ K (Peebles 1971), so we find

$$\rho_{HI} \lesssim 3 \times 10^{-28} \text{ kg m}^{-3}.$$

4.10. Lyman-α Absorption

In principle, the presence of intergalactic neutral hydrogen should be revealed by any number of possible absorption lines; in practice only the strongest lines would provide any possibility of detection, and this implies absorption from the ground state, since this will be the most populated state. Apart from the hyperfine radio lines, this leaves the Lyman series starting with Lyman-α at 1215 Å in the ultraviolet region of the spectrum, which cannot be observed from below the atmosphere. However, for quasars with redshifts greater than about 2, any ultraviolet absorption features should be shifted into the visible part of the spectrum. In the spectra of most high-redshift quasars no strong absorption edge is found at 1215 Å, so the optical depth must be less than unity, and this provides a limit on the amount of intergalactic neutral hydrogen. Less spectacular absorption features are in fact found in many quasars, from which one may hope in due course to learn more details of the composition of the intergalactic medium.

The calculation of Lyman-α absorption is similar to that for 21 cm absorption except that here the majority of atoms are in (one of the levels of) the ground state. Hence the corresponding spin temperature, T_s, satisfies $kT_s \ll h\nu_{Ly-\alpha}$, and the emission coefficient now contains the Boltzmann factor $\exp(-h\nu_{Ly-\alpha}/kT_s)$. The ratio of statistical weights of the 1S and 2P levels involved in the transition is $\frac{1}{3}$, so the emission coefficient is

$$j \approx \frac{1}{4\pi} [3 \exp(-h\nu_\alpha/kT_s)] (h\nu_\alpha) A_{21} n,$$

where A_{21} is the 2P $\rightarrow$ 1S transition rate (or Einstein coefficient), and we have written $\nu_\alpha = 2.5 \times 10^{15}$ Hz for the frequency of the Lyman-α line. Thus the mean

75

opacity over the line is

$$\bar{\kappa} = \frac{1}{\Delta\nu} \int_{\nu_\alpha - \Delta\nu/2}^{\nu_\alpha + \Delta\nu/2} \kappa_\nu \, d\nu = \frac{1}{\Delta\nu} \int j_\nu/B_\nu \, d\nu$$

$$\approx j/B_{\nu_\alpha} \Delta\nu = \frac{3}{8\pi} \left(\frac{c}{\nu_\alpha}\right)^2 \frac{nA_{21}}{\Delta\nu} [1 - \exp(-h\nu_\alpha/kT_s)].$$

Since we have $kT_s \ll h\nu_\alpha$, we can neglect the exponential term in this expression (i.e. there is no correction for stimulated emission). Then, exactly as before, the optical depth at frequency ν becomes

$$\tau_\nu = \frac{3}{8\pi} n_{HI}(t) A_{21} \left(\frac{c}{\nu_\alpha}\right)^3 \frac{1}{H}, \tag{4.10.1}$$

where H and n_{HI} are to be evaluated at the time when the redshifted frequency, ν, that one is actually looking at would have been at ν_α in the local rest frame, and therefore capable of being absorbed. At this time we have a redshift given by

$$1 + z_\alpha = \nu_\alpha/\nu.$$

At this point there is a crucial difference between this test and the search for 21 cm absorption. For the Lyman-α absorption occurs at large distances in the past and is redshifted into the visible, whereas the absorption at 21 cm occurs locally. To use equation (4.10.1), we therefore need to know the intergalactic density and Hubble constant at z_α. This requires a cosmological model for the change of H with time, or, equivalently, with redshift, and so a theoretical framework. This we shall provide in Chapters 5 and 6. Here we can use some simple arguments to fix orders of magnitude; in the context of our later discussion, these results here correspond to the case $q_0 = 0$.

We can argue as follows: (i) H is of order t^{-1} since the age of the Universe is of order H^{-1}. (ii) From Hubble's law, $-\Delta\nu/\nu = H\Delta t \sim \Delta t/t$, so $\nu \propto 1/t$. (iii) $\dot{R}/R = H \sim 1/t$ implies $R \propto t$, and hence $(1+z) \propto 1/t$ is the relation between the observed redshift of an object and the time at which the signals received now were emitted. It follows also that $H \propto (1+z)$. (iv) Finally, $n_{HI}(t) \propto R^{-3} \propto t^{-3} \propto (1+z)^3$ gives the redshift dependence of the density in this model. Hence, in terms of present quantities, we have

$$\tau\left(\frac{\nu_\alpha}{1+z}\right) = \frac{3}{8\pi} (n_{HI})_0 A_{21} \left(\frac{c}{\nu_\alpha}\right)^3 \frac{(1+z_\alpha)^2}{H_0}.$$

Since the relevant quasars have $z \sim 2$, the condition $\tau < 1$ implies

$$n_{HI} \lesssim \frac{4\pi}{9} \left(\frac{\nu_\alpha}{c}\right)^3 \frac{H_0}{A_{21}} \approx 2.4 \times 10^{-5} \, \text{m}^{-3}.$$

or

$$\rho_{HI} \lesssim 4 \times 10^{-32} h \, \text{kg m}^{-3}.$$

The limit for a more realistic cosmological model is even tighter than this.

This limit is much more restrictive than those provided by the 21 cm tests. In fact, the allowed density is so small that the correct conclusion is assumed to be not that there is essentially no intergalactic medium, since this would require galaxy formation to be almost 100% efficient in its use of the available matter, but that the neutral hydrogen content is low because the gas is kept ionised.

4.11. Tests for an Intergalactic Plasma

The intergalactic medium may be ionised either as a result of photoionisation by a source of ultraviolet photons or by collisions between neutral atoms and electrons of sufficient energy. In the former case the plasma temperature will be around 10^4 K, since above this temperature electrons lose energy by recombining with increasing effectiveness, and this loss rate cannot be balanced by any typical ionising continuum. In principle, there is a somewhat contrived intermediate case in which the free-free emission from the gas at somewhat below 10^6 K is itself sufficient to keep the medium photoionised. A more likely alternative is a gas above 10^6 K, when collisional ionisation is sufficient to reduce the density of neutral matter to below the observed limits, even if the mass in the intergalactic medium is an order of magnitude higher than the mass in galaxies. In any case, it is necessary to postulate a source of energy input into the gas since otherwise, left to itself, the intergalactic gas would by now be neutral at a temperature well below that of the microwave background! We shall consider two tests, one appropriate to the low-temperature case, and one to the high-temperature case, which each yield upper limits for the density. Further, more exotic possibilities are summarised, for example, by Rees *et al* (1974).

4.12. Lyman-α Emission

A photoionised gas emits radiation as the free electrons and ions recombine. If such radiation is produced in the intergalactic medium, the low density of neutral hydrogen found in §4.10 will allow almost all of it to propagate to us without absorption. Taking into account the redshift of the emitted radiation, we would expect a step in brightness longward of the rest frame wavelength of Lyman-α at 1215 Å, just as we found for 21 cm emission. As in §4.8, the emission coefficient is

$$ j = \frac{1}{4\pi} \cdot n_e n_p h\nu_\alpha \alpha_{2P}, $$

where α_{2P} is the recombination rate to the 2P level for electrons recombining on to a single proton in unit volume, both directly and by cascading down from higher levels. In the steady state, α_{2P} equals the transition rate 2P → 1S. The quantities n_e and n_p are the number densities of electrons and protons, and we can set $n_e \approx n_p = n_{HII}$ for an essentially pure hydrogen medium. The step in brightness is therefore given by

$$\Delta i_{\nu_\alpha} \, d\nu = j \, ds = \frac{jc}{\nu_\alpha H} \, d\nu$$

at $\nu_\alpha = 2.5 \times 10^{15}$ Hz, or

$$\Delta i_{\nu_\alpha} = \frac{\alpha_{2P}(n_{H\,II})^2 hc}{4\pi H_0}.$$

The observed step, estimated from a measurement of the ultraviolet background near 1215 Å, is $\Delta i_\nu \lesssim 10^{-24}$ W m^{-2} sr^{-1} Hz^{-1} (§3.1). The recombination coefficient depends weakly on temperature; for $T \approx 10^4$ K, we have $\alpha_{2P} \approx 13.5 \times 10^{-20}$ m^3 s^{-1}, and, consequently, a rather weak limit

$$\rho_{H\,II} \lesssim 2 \times 10^{-26} h^{1/2} \text{ kg m}^{-3}.$$

However, there are problems with this picture. Photoionisation of the inter-galactic gas is certainly not important now, since there is no strong ultraviolet background. Recombinations are therefore not followed by subsequent ionisation, and any emission must therefore occur at the expense of an increase in the neutral hydrogen content. If one postulates that photoionisation occurred too long ago, the plasma has had time to recombine and there is a conflict with the limits on the neutral hydrogen density. But the absence of evidence for strong intergalactic absorption by neutral hydrogen in the spectra of quasars at $z \sim 2$ (§4.9) tells us that the gas must be ionised at least back to this redshift. It must have at least partially recombined by the present epoch. On the other hand, a flux of ultra-violet radiation sufficient to ionise a dense intergalactic medium as recently as a redshift of 2 would probably conflict with direct measurements of the present ultraviolet background. Therefore, even though direct observations of Lyman-α emission do not rule it out, it is difficult to arrange for the presence of a low-temperature plasma of significant density.

4.13. X-ray Emission

For collisional ionisation of the intergalactic plasma, temperatures in excess of 10^6 K are required. X-ray emission from the gas will then be important; indeed, attempts have been made to account for the observed isotropic x-ray background on the basis of thermal bremsstrahlung from an intergalactic medium with appropriate temperature and density. For the present discussion we can investigate the limits imposed by the condition that the intergalactic x-ray emission should not exceed the observed x-ray background. This argument is independent of the origin of the background.

The limits obtained depend on the details of the model assumed for the tem-perature history of the gas. For simplicity we assume that the gas is suddenly heated to x-ray temperatures, say 10^8 K, at $z \sim 2$, and then cools in such a way that

$T \propto (1+z)$. (In fact, the result depends only weakly on this assumption.) The emission coefficient for bremsstrahlung from a thermal plasma (in which the radiation arises from electrons accelerated in the electric fields of the ions), is

$$j_\nu \, d\nu \approx 5 \times 10^{-40} [n_e(t)]^2 T^{1/2} \exp(-h\nu/kT) \, d\nu \ \text{W m}^{-3} \, \text{sr}^{-1}$$

at the point of emission at time t. In terms of the observed frequency, the contemporary density, n_e, and the temperature, T_0, this is

$$(1+z) j_{\nu_0(1+z)} \, d\nu_0 \approx 5 \times 10^{-40} n_e^2 T_0^{-1/2} \exp(-h\nu_0/kT_0)(1+z)^{13/2} \, d\nu_0$$

using $\nu = \nu_0(1+z)$ and $n_e(t) = n_e(1+z)^3$.

This has to be integrated along the line of sight to give the observed intensity, which means that we need a cosmological model to tell us, in physical terms, how much time the Universe spent at each redshift — or, more mathematically, to convert the integral from an integration over t to an integration over z. To provide an order of magnitude estimate, we can again use the model with $H \propto 1/t$, as in §4.10. There we found $t^{-1} = (1+z)H_0$, from which

$$i_{\nu_0} \, d\nu_0 = (1+z)^{-1} \int j_{\nu_0(1+z)} c \, dt \, d\nu_0$$

$$\approx 5 \times 10^{-40} n_e^2 T_0^{-1/2} \exp(-h\nu_0/kT_0) \frac{c}{H_0} \int_0^2 (1+z)^{7/2} \, dz.$$

If $h\nu_0 \ll kT_0$,

$$i_{\nu_0} \sim 100 \times 5 \times 10^{-40} (c/H_0) \, n_e^2 \, T_0^{-1/2}.$$

The observed value of i_{ν_0} at $\nu_0 \sim 10^{18}$ Hz is $i_{\nu_0} \approx 10^{-27} \text{W m}^{-2} \, \text{sr}^{-1} \, \text{Hz}^{-1}$. If $T_0 \approx 10^7$ K, this yields

$$\rho_{\text{H\,II}} \lesssim 10^{-28} h^{1/2} \ \text{kg m}^{-3}.$$

The problem of x-ray emission has been investigated in much more detail than given here, with the result that a high-density Universe cannot be ruled out from the present x-ray data. Thus our result, while reasonable as an order of magnitude for the specified model, is not to be taken too seriously. The discussion does, however, illustrate the method of investigation.

In this case the heating and consequent ionisation of the intergalactic medium might be accomplished by dissipation of the bulk kinetic energy of the gas, or by low-energy cosmic rays, or by relativistic particles escaping from quasars.

4.14. Cosmic Rays

The Earth is bathed not only in radiation, but also in a flux of energetic particles which are called cosmic rays. Energies up to 10^{20} eV per particle have been detected, corresponding to protons moving to within about one part in 10^{22} of the speed of light. For comparison, the largest laboratory accelerators achieve energies of order

10^{12} eV. Nevertheless, the number of particles involved is relatively small, so the contribution of cosmic rays to the energy density of the Universe is negligible. Within the Galaxy it is approximately the same as that contained in starlight and magnetic fields (10^{-13} J m^{-3}), and it is certainly no higher in intergalactic space.

Protons are the main constituents of cosmic rays, but alpha-particles are also present in roughly the proportion corresponding to the universal abundance ratio of helium to hydrogen (about 10% by number). There is a large over-abundance of light nuclei (Li, Be, B), and heavy nuclei are also found. Electrons are present in cosmic rays with an abundance of about 1% of the number of protons of the same energy. The feature of key cosmological significance in the composition of cosmic rays is that not one single antiparticle has ever been detected.

At least for higher energies, where the contribution from the Sun and the effect of the Earth's magnetic field become unimportant, the cosmic ray flux is isotropic with upper limits for the variation in intensity, $\delta I/I$, ranging from about 10^{-3} at 10^{14} eV to 0.1 at 10^{19} eV. As always, this high isotropy can be taken to imply that the sources are not local. It does not, however, imply that the cosmic rays must have an extragalactic origin, although such a possibility would provide a natural explanation of the isotropy and is not ruled out. This is because the orbits of the charged cosmic ray particles are bent by the Galactic magnetic field sufficiently to destroy anisotropies at all but the highest energies. Indeed, the uniformity of the diffuse Galactic radio emission (§3.1) is evidence that the cosmic rays fill the Galaxy.

The energy distribution amongst cosmic ray protons above about 10^9 eV is described by a differential energy spectrum $I_E = 2.5 \times 10^{18} E^{-1.6}$ eV m^{-2} s^{-1} sr^{-1} eV^{-1}, steepening somewhat around 10^{15} eV, and regaining its original slope again at about 10^{17} eV. This may be an indication of two (or more) components to the flux; in particular, the break at 10^{17} eV may indicate an extragalactic component which would then contribute about 1% of the flux at lower energies.

If neutrinos are confirmed to be massive particles, as has been recently suggested, then they too should be included as a component of cosmic rays. If they are present in the abundance expected from a hot big bang, their density should be sufficient to cause the Universe to recollapse.

4.15. Antimatter

Seen from a distance, an antimatter star is indistinguishable from a star of matter, an antimatter Moon from one made of matter. From one point of view the Ranger spacecraft may be regarded as (unnecessarily sophisticated) lunar antimatter detectors. The failure to observe spontaneous annihilation of these detectors can be taken as conclusive proof that the Moon is made of matter rather than anti-matter. This method is not at present capable of significant extension. Since, then, we cannot go to the Universe, we must wait for it to come to us. It is here that the

80

cosmic rays play an important role, providing us with our only direct evidence on the antimatter content of the Universe.

Upper limits for the fraction of antiparticles in cosmic rays range from somewhat less than 10% to 3×10^{-4} (Steigman 1973), as estimated from failures to detect antiprotons, anti-alpha particles and antinuclei at various energies. Antiprotons are expected at some level since they can be produced in collisions between cosmic rays and the interstellar medium, but the detection of a single anti-alpha particle or antinucleus would be strong evidence for the existence of domains of antimatter in the Universe.

The conclusions to be drawn from the present upper limits depend on from where the cosmic rays are assumed to originate. If they are universal then no more than 10%, and possibly less than 3×10^{-4}, of the Universe is made of antimatter. On the other hand, although we know that the cosmic rays almost certainly fill the Galaxy, it is conceivable that those we detect at present originate in a region near the Sun (from say 100 pc), and diffuse only slowly through the Galaxy. In this case more distant regions of the Galaxy might consist of antimatter. This possibility is in fact ruled out by other indirect evidence, since the gamma-rays produced by matter-antimatter annihilation would exceed the upper limits on the Galactic gamma-ray background.

The most conservative estimate for the extragalactic component of cosmic rays would appear to arise from the assumption of a reasonable leakage rate from galaxies. If all the cosmic ray particles at present in the Galaxy were to have been here for a significant fraction of a galactic lifetime then they would have had time to interact with heavy elements in the intergalactic gas. This would have produced large amounts of the light elements (Li, Be, B) by 'spallation', in which process the light nuclei are chipped off the heavy ones. We know that the cosmic ray density at the Earth has been virtually unchanged for about 3×10^9 years from observations of tracks in rocks and meteorites. If we require that no more of the light elements than are seen be produced in this way, we cannot allow the cosmic rays to shuttle across the Galaxy for so long a time. The resolution is to allow them to leak out in about 10^6 years, so that each galaxy populates intergalactic space with about 3×10^3 times as many cosmic ray particles as there are in the Galaxy. This allows one to calculate that about 3×10^{-4} of the observed cosmic ray flux could come from the Virgo supercluster. If half of the Supercluster were antimatter we would expect a proportion of 1.5×10^{-4} of antiparticles in the cosmic ray flux. Thus the available upper limits are not sufficiently stringent to rule out this most conservative possibility directly, although this conclusion does require that there should be no significant output of relativistic particles from active antigalaxies or radio antigalaxies. An analogous argument does, however, show that the Local Group must be predominantly made of matter.

A certain amount of indirect evidence for the absence of a significant quantity of antimatter comes from limits on the gamma-ray background. Since gamma-rays would be produced by particle–antiparticle annihilation it is possible to deduce that, if an intergalactic medium of reasonable density exists, then it can contain equal amounts of matter and antimatter only if these are in separate regions of a size of the order of the observable Universe. It is also possible that there is essentially no gas between rich clusters of galaxies and that half of those are made of matter and half of antimatter. The problem with these pictures is that it is difficult to see how to achieve ultra-efficient separation of matter and antimatter or to avoid an intergalactic medium. While the picture of a Universe symmetric between matter and antimatter is therefore possible, it seems more likely that this is a symmetry of which the Universe does not avail itself.

4.16. Matter and Radiation

If the Universe is not symmetric between matter and antimatter, then the excess of the one over the other is an arbitrary parameter apparently to be specified along with the initial conditions of this Universe. This brings us to one of the deeper mysteries of the hot big-bang model.

The ratio of the amount of radiation to the density of matter in the Universe can be characterised by a dimensionless constant, s/k, which is the ratio of entropy per unit volume in the radiation field to the density of matter at any time. From §3.6 this is

$$\frac{s}{k} = \frac{4/3 \, aT^3}{nk},$$

with Boltzmann's constant, k, introduced simply to make s/k dimensionless. We assume that the temperature of the radiation satisfies $T \propto 1/R$ as the Universe expands, even though we have proved this at present only for a pure radiation universe (§3.6; see Chapter 8). It follows that s/k is constant in time, since we know that $n \propto R^{-3}$. For a 'low-density' Universe with $\rho = 10^{-27} \, \text{kg m}^{-3}$ now, and $T(\text{now}) = 2.7$ K, we find $s/k \sim 10^9$. According to the big-bang model this is another constant of Nature. We can see its relation to the antimatter content if we interpret it in a slightly different way.

At early times in the evolution of the Universe, the hot big-bang theory postulates a phase of thermal equilibrium between matter and radiation. In thermal equilibrium at temperature T, an average photon has energy of order of magnitude kT. The number of photons in unit volume of space is therefore of order $aT^4/kT = T^3/k$, since the energy density is aT^4. Thus s/k can be interpreted as the number of photons per material particle. At sufficiently early times, for temperatures in excess of 10^{12} K, photons have sufficient energy to produce proton-antiproton pairs. The condition of thermal equilibrium then requires a universal density of antiprotons (and indeed of antineutrons, positrons, etc), such that their rates of creation and annihilation are equal. Strictly, this means that we should consider s/k to represent the number of photons per unit excess of particles over antiparticles, or, expressed more conventionally, as the number of photons per baryon. For each pair of photons at this early stage there was approximately one particle–antiparticle pair, so if we count up the number of particles and anti-particles we find, to order of magnitude, that for every 10^9 antiprotons we must have $10^9 + 1$ protons. Running the system forwards in time again from this point, we see 10^9 particle–antiparticle pairs annihilate to contribute to the microwave background for every particle that is left over to make the stars, the planets and Man.

From this point of view, the material world appears as an almost ludicrously minute imbalance, an apparent cosmic accident, in the equality of matter and antimatter, which would otherwise have led to an endless uniform sea of black-body photons. It is not hard to feel the need for a less arbitrary machinery of creation than is provided by this standard hot big-bang model. This is the motivation behind attempts to allow a symmetric universe through the efficient separation of matter and antimatter domains (Omnes 1972). An alternative approach envisages the generation of 10^9 photons per baryon by dissipative processes radiating the energy of a reasonable spectrum of fluctuations and inhomogeneities in cold matter (Zeldovich 1972, Rees 1972). More recently, the grand unified theories of particle interactions have appeared to offer a possible explanation (see § 12.2).

5.   The General Theory
of Relativity

5.1. The Laws of Physics

Our observation of the Universe is limited in practice by our ability to collect and interpret data. Even in principle we have access only to information that happens to have intersected the orbit of a civilised Earth. We can look at only a frozen image of the Coma cluster, at only an instant in the history of each galaxy. The quasar 3C 273 appears as it was 1.6 billion years of evolution ago. For the most distant objects we cannot even assign an age without some model for the dynamics of the whole Universe. Furthermore, we cannot unravel the origins of what we see by running the Universe backwards in time, for the formation of structure involves irreversible processes of dissipation which cannot be unwound. Thus the systematisation of the observational record requires the theoretical prediction of evolutionary sequences. A good example of this is the microwave background, which we have interpreted in terms of an unobserved hot dense phase of the Universe. In this case it is the theory of the evolution of radiation in an expanding universe that tells us, except at a superficial level, what it is we are seeing. These remarks are not unique to cosmology, of course. No amount of observing of stars or experimenting with atoms can tell us what they are.

The theory of the dynamical and physical development of the Universe forms the subject of the subsequent chapters. A certain amount of speculation, beyond the realms of the physics laboratory is inevitable here. There are two ways in which this can be limited; by conservation and completeness. We aim to keep to a minimum the extrapolation from what we know to be the case by experiment, and we try to avoid the introduction of particular *ad hoc* hypotheses for individual phenomena and aim instead at a single overall picture.

In order to begin theoretical cosmology, therefore, we assume that the physics which we learn on Earth will remain valid for all time and can be applied throughout the Universe. It is possible that certain quantities we interpret as constants of Nature might be slowly varying as functions of the age of the Universe, or might be dependent on environmental conditions. For example, the strength of gravity could be changing, or the proton might be unstable near a large agglomeration of matter. Experiments can put limits on such variations, since we do survey a finite, albeit small portion of the Universe in the course of the Earth's motion through

84

space. But experiments can never entirely rule out small variations which could be significant on a cosmic scale, so our assumption cannot be checked with the relevant precision.

A related but distinct assumption concerns the scope of the applicability of known physics. We assume that it makes sense to treat the Universe as a whole as subject to dynamical laws. Thus we may postulate that the laws of gravity are valid in the same way for every planetary system throughout the Universe, but this is distinct from the assumption that these same laws of gravity can be applied on the scale of the visible Universe. In particular, there is a problem, to which we shall return in Chapter 10, of formulating a distinction between physical laws and initial conditions when considering a system of which there is only one example. In order to explain any property of the Universe we can simply postulate it as an appropriate initial condition and assert this as a physical law. One can therefore argue that it is not known exactly what is meant by 'the laws of physics' in this context.

We eliminate all of these problems in the first instance by ignoring them. The *fundamental speculation of theoretical cosmology* is that the terrestrial laws of physics are adequate for the construction of models of the Universe. In this chapter we shall be concerned to establish the appropriate theory of gravity for application to the Universe in Chapter 6.

5.2. The Role of Gravity

The motions of bodies are controlled by the forces between them. We recognise four types of force between particles, although the four types are not necessarily entirely distinct at a fundamental level. In order of decreasing strength the forces are the 'strong' or 'nuclear' force, which is responsible for the stability of the atomic nucleus, the electromagnetic force, the 'weak interaction', which is responsible for beta radioactive decay, and gravity. If we assign unit strength to the strong force then the others have strengths of order $1/137$, 10^{-16} and 10^{-40}, respectively. The strong and weak interactions are effective over only very short ranges and are therefore irrelevant on a cosmological scale; they were important only in the very early stages of the Universe when particles were squeezed very tightly together. The weak interaction, for example, was important on a small scale in maintaining equilibrium between neutrons and protons when the Universe was less than about one second old (Chapter 8). The electromagnetic force is long range, but because on average, cosmic bodies appear to be electrically neutral the net electromagnetic force between them is zero. However, it played a role in the early stages, when matter was broken up into its electrically charged constituents, in maintaining an equilibrium between radiation and matter. But the hypothesis of overall charge neutrality ensures that electromagnetism had no large-scale dynamical role. This leaves gravity, the weakest of the forces, to control the large-scale dynamics of matter.

5.3. Newtonian Gravity

The Newtonian theory of gravity provides an adequate description of gravitational phenomena from the terrestrial scale up to the scale of clusters of galaxies. If one observes accurately the motion of Mercury, or the bending of starlight by the gravitational field of the Sun, then small discrepancies appear between the observations and the predictions of the theory. Applied to the Universe as a whole, however, Newton's theory was thought to give not merely results at variance with observation, but no consistent results at all. Suppose first that the Universe were a finite system of stars or galaxies. Such a system cannot remain static, but must collapse under the mutual gravitational attraction of its constituents in a finite time. Because the Universe was supposed to be static, this appeared to rule out the possibility of its being finite. Suppose then that the Universe were infinite and static. To determine the effects of gravity we need to know the gravitational potential ϕ at each point due to the constituent matter; this is determined by solving the Poisson equation

$$\nabla^2 \phi = 4\pi G\rho, \tag{5.3.1}$$

where ρ is the density of matter. If the gravitational potential ϕ is not constant on average, matter will tend to agglomerate in regions of low potential and high density. A static situation is obtained only if ϕ is a constant. But then $\nabla^2 \phi = 0$, and hence $\rho = 0$; this solution would therefore appear to be of no physical relevance.

It appears that one must either abandon attempts to understand the Universe on the basis of physical laws, or the law of gravity must be modified. One suggested modification (see North 1965) is to amend the Universe square law such that the force on a mass m due to a mass M at distance r becomes

$$F = GmM \exp(-\Lambda^{1/2} r)/r^2. \tag{5.3.2}$$

The introduction of the new constant of nature, Λ, produces an effective cut-off in the gravitational force at separations of the order of $\Lambda^{-1/2}$. If Λ is of the order of $10^{-50}\,\mathrm{m}^{-2}$, the cut-off occurs at the radius of the observable Universe, and this would not sensibly affect planetary orbits.

However, the argument that Newtonian theory led to inconsistencies was based on the assumption of a static Universe with an infinite past. It can be looked at as simply the gravitational analogue of Olbers' paradox. A way out of the difficulty is to postulate an expanding Universe with a finite past history. A spatially finite expanding model then yields a consistent solution provided it is sufficiently large that its edge is beyond the limit of observation. Also, the Newtonian theory can be formulated in such a way as to allow an infinite expanding model, but in the absence of observational evidence, not even Einstein was bold enough to predict a non-static Universe on the basis of such considerations.

5.4. Gravity and Special Relativity

A more serious difficulty with the application of the Newtonian theory of gravity in cosmology is the failure of the theory to be compatible with special relativity. One might suspect special relativistic effects to be important since some quasars are observed to exhibit redshifts, z, in excess of unity. This is incompatible with a Newtonian interpretation of the Doppler effect, since one would obtain velocities, $v = cz$, in excess of that of light. The special relativistic Doppler formula $1 + z = (c + v)/(c - v)^{1/2}$ always leads to sub-luminal velocities for objects with arbitrarily large redshifts, and is at least consistent. In fact we shall find that the strict special relativistic interpretation is also inadequate. Nevertheless, at the theoretical edge of the visible Universe we expect at least in principle to see bodies apparently receding with the speed of light. In these circumstances one would not expect to construct a successful model by neglecting relativity entirely.

An alternative argument for the importance of relativistic effects can be stated in terms of energies. The limit of the validity of Newtonian dynamics, and hence the point at which relativistic considerations are required, occurs when the energy of a system becomes comparable with the rest mass energy of the matter. The gravitational potential energy of a system of mass M in radius R is approximately $- GM^2/R$. The magnitude of the gravitational energy of the Sun, $GM_\odot^2/R_\odot \sim 10^{40}$ J, is much less than its rest mass energy, $M_\odot c^2 \sim 10^{47}$ J, and so relativistic effects are negligible. However, on the scale of the visible Universe at the present epoch we find

$$GM^2/R \sim Mc^2 \qquad (5.4.1)$$

for a density of matter of order 10^{-26} kg m^{-3}. We shall certainly want to consider densities of this order.

How is it then that Newtonian gravity comes into conflict with special relativity? Clearly, the inverse square law of gravitational attraction cannot be valid in all frames of reference connected by Lorentz transformations. It follows that observers in uniform relative motion who obtain identical results for electrodynamic experiments would predict different gravitational behaviour according to the Newtonian theory. This would enable us to pick out a single privileged rest frame by means of local physical experiments, namely, that unique frame in which bodies are attracted gravitationally exactly as the inverse square of the distance. This contradicts the Principle of Relativity, according to which uniform motion cannot be detected by local experiments.

Despite this, we know that electrically charged bodies at rest interact through the electrostatic inverse square law force. How can it be that the electrostatic inverse square law is compatible with special relativity whereas the same law of gravity is not? The crucial point is that the electrostatic force is only part of the way in which electrically charged bodies interact; in general, there is in addition a magnetic component to the force which depends on the motion of the bodies. Different observers ascribe the same observed dynamics to the influence of different combinations of electric and magnetic fields. On the other hand, according to the Newtonian theory, there is only a static gravitational interaction. Clearly the question arises as to whether one can bring about compatibility between special relativity and gravity by adding to the gravitational force extra components depending on the motion of bodies. In a sense, the general theory of relativity, which is the name Einstein gave to his theory of gravity, can be thought of as achieving essentially just this. In Newtonian gravity the force exerted by a rotating sphere on a nearby body is exactly the same radial force as that exerted by the sphere at rest. In Einstein's theory the rotating body exerts a small additional transverse component.

This way of describing Einstein's theory is adequate when gravitational forces are relatively weak, and can even be made to yield quantitative results. In the case of strong fields – in particular, when equation (5.4.1) is satisfied, which is the case of interest in cosmology – the foregoing picture is incomplete, for there is an important difference between the electromagnetic and gravitational forces. Only electrically charged entities can contribute to or respond to electromagnetic fields, and not all objects are charged. In particular, electromagnetic radiation does not carry charge so electromagnetic energy does not itself generate an electromagnetic field. On the other hand, all energy is equivalent to mass, and all mass contributes to gravity. Therefore the potential energy of the gravitational interaction itself contributes to the gravitational force between bodies; in this sense the gravitational field has weight! Indeed, one can read equation (5.4.1) as an expression for the mass to be attributed to the gravitational energy of the Universe. As long as the equivalent mass of gravity is negligible the analogy with electrodynamics is satisfactory. Otherwise we are forced to adopt a more general point of view, which is appropriate for both weak and strong gravitational fields. We have to give up the idea of gravity as an ordinary force represented by a vector field in a space with pre-assigned geometrical properties, and describe it instead in terms of the changing geometrical properties of space–time; in particular, gravitational effects arise as an aspect of the way in which we measure time. In the following two sections we shall see how this apparently strange idea arises in a natural way from some simple physical considerations.

5.5. The Principle of Equivalence in Newtonian Physics

The general theory of relativity is based on a simple physical idea which is referred to as the Einstein Principle of Equivalence (Will 1974). The complications of the theory arise only in the technical elaboration required to turn this idea into a workable tool for the purpose of calculation of general results. We shall be able to avoid most of these complications by restricting the discussion to just the particularly simple cosmological models which appear to provide realistic models for our actual Universe.

Let us neglect for the moment the requirement that a realistic theory of gravity be compatible with special relativity. Consideration of the Principle of Equivalence will lead us to conclude that, even so, Newtonian gravitational theory is conceptually ill-founded, despite the fact that it can be made to yield adequate numerical predictions within its domain of validity. From this point of view Einstein's development of general relativity can be taken as a critique of Newton's formulation of his theory of gravity. When this criticism is absorbed into a modified Newtonian theory, the result bears a strong resemblance to the general theory of relativity. The point therefore is that we can discuss what form a theory of gravity should take without mentioning special relativity. This is the subject of the present section. Only subsequently (in §5.6) do we take account of the requirements of relativity. It will turn out to be possible to do this in a natural and straightforward way.

First we need to recall the notion of a *test body*. In order to detect the presence and strength of a gravitational field we introduce a massive body and observe its motion. In principle the presence of this extra mass disturbs the system we are trying to measure. We therefore imagine that the extra body has a sufficiently small mass that this disturbance is less than the limit of accuracy of the measurement and can therefore be neglected. In classical (i.e. non-quantum) physics this is always possible. If this is the case we refer to the probe as a test body. Of course, we require in addition that a test body should be geometrically small in order that we may consider it as responding to the gravitational field at a point. In practice this means that the test body must be smaller than the distance over which variations in the strength of gravity are significant.

We can now state a form of the Principle of Equivalence:

The Galilean Principle of Equivalence states that all test bodies fall with equal acceleration in a gravitational field, independently of their internal structure and composition, provided there are no other disturbing forces.

This is, of course, the statement that, according to legend, Galileo is supposed to have verified by dropping objects from the tower of Pisa. The specification of the absence of disturbing forces is necessary since, for example, a charged particle would fall in the Earth's magnetic field differently from an uncharged one.

The first accurate experiment to test the Principle of Equivalence was carried out by Baron Eötvös in 1890. In this experiment two weights are hung from the ends of a beam which is suspended so that it can rotate freely. The arrangement is such that any inequality in the acceleration of the weights will cause a systematic rotation of the beam. By balancing wood and platinum spheres, Eötvös was able to show that in this case there was no anomalous (composition-dependent) gravitational force of magnitude greater than one part in 10^9 of the total force. Dicke (1964) and a team at Princeton have searched for a failure of the Principle of Equivalence in the gravitational field of the Sun. They used a balance similar in principle to that of Eötvös but looked for a torque with a period of 24 h, which is just the orbital period of the Sun around the apparatus. The periodicity of the effect, if present, enhanced the sensitivity of detection, and this enabled Dicke to set a limit of one part in 10^{11} for any anomalous difference between the force on aluminium and gold weights. These materials were chosen because they have very different internal properties in their nuclear and atomic configurations. More recently, Braginsky and Panov (1972) improved the accuracy to one part in 10^{12} in a similar experiment. We shall assume that the Principle is satisfied exactly.

Galileo's Principle is incorporated into the Newtonian theory in a very simple way. To see this we must first distinguish between three types of mass that occur in the theory. These are the inertial mass, the passive gravitational mass and the active gravitational mass. The inertial mass, m_i, measures the inertial resistance of a body to acceleration by an imposed force, and is the mass that enters Newton's second law as mass times acceleration. The passive gravitational mass, m_p, measures the response of a body to a gravitational field; if the gravitational potential is ϕ, the force on the body is $-m_p \nabla \phi$. The active gravitational mass measures the capacity of a body to produce a gravitational field, so the potential a distance r from a spherical body of mass m_a is $-Gm_a/r$. In this language one can think of a test body as a hypothetical object with no active gravitational mass, so that it does not itself produce a gravitational field, but possessing a passive gravitational mass, so that it can respond to an existing field. We are used to thinking of the different types of mass in Newtonian theory as identical but, strictly speaking, they are not. The numerical equality of m_p and m_a can be shown to be a consequence of Newton's third law, that action and reaction balance. The numerical equality of m_i and m_p is equivalent to the Galilean Principle of Equivalence. For Newton's second law gives $m_i a = -m_p \nabla \phi \equiv m_p g$, and the acceleration, a, due to gravity is the same for all bodies, $a = g$, if and only if $m_i = m_p$. In this way the Principle of Equivalence is satisfied at the cost of introducing two different concepts of mass which just happen to be numerically equal.

However, the inadequacies of Newtonian gravity are not merely aesthetic. The Principle of Equivalence actually undermines the conceptual basis on which the theory in its standard form is founded. According to Newton, in order to discuss a dynamical problem, we first identify a frame of reference which may be taken to be at rest and which is suitable for application of the second law of dynamics. Such a frame is called an inertial frame. For many purposes, and for the sake of argument here, the surface of the Earth can be taken to provide such a frame, although for greater accuracy the rotation and motion of the Earth must be taken into account. We then observe that certain bodies, apparently subject to no external forces in this inertial frame, do *not* move in straight lines at constant speeds but fall to the Earth with a common acceleration independent of the masses of the bodies. We attribute this phenomenon to a new force which we call gravity. However, for conceptual propriety we should now go back and check that our frame of reference really is an inertial frame for the purpose of describing this new force. If we were dealing with, say, the electromagnetic force, in the absence of gravity, this would be straightforward: we simply observe that electrically neutral bodies move with constant velocities to confirm that we are indeed in an inertial frame of reference. But we cannot turn off gravity in this way since the effects of gravity are independent of mass. What experiment can we perform to tell us that we are at rest? One would like to answer that all we have to do is to look at the laboratory floor, but this is not adequate. Looking at floors is not an experiment within the context of dynamical theory, and merely confirms a prejudice sanctioned by usage; for we already know that the 'fixed' floor is accelerating with the Earth. In fact, it follows from the Principle of Equivalence that no dynamical experiment at all can be used to answer the question.

Consider, for example, an enclosed laboratory accelerated at a constant rate, g, say by rocket motors, in the effectively gravity-free environment of outer space. Any objects dropped by an experimenter in the laboratory will be seen to fall with an acceleration, $-g$, which is independent of their composition and mass. On the basis of the Galilean Equivalence Principle he may conclude that he is at rest in a gravitational field. The conclusion can be extended to any experiment. Conversely, an experimenter on Earth sees objects fall exactly as they would in an accelerated laboratory in space. There is no way of establishing that one is at rest in a gravitational field since the Equivalence Principle implies that this state does not differ from one of uniform acceleration in gravity-free space.

In practice we avoid this difficulty by judging our state of motion with respect to the distant stars or galaxies. These, it is claimed, provide an absolute standard of rest — and so they do, in so far as they enable us to extract correct results from the Newtonian theory. But Einstein noticed that this was an arbitrary solution to the problem because, according to the theory itself, nothing at all would be changed in our local environment if all the stars were to be removed! By cheating in this way, one is missing the real point. For if one cannot *tell* the difference, by means of local experiments within the scope of the theory, between an accelerated frame of reference and the presence of gravity, then to this extent there *is* no difference. The restriction to local experiments is crucial here and will be explained in due course; for the present it rules out the use of systems of test bodies which are large compared to the distance over which the gravitational field produces variable accelerations, and observations of the distant stars.

The reason for the name 'Equivalence Principle' for Galileo's observation is now clear, since it implies the equivalence of gravity and acceleration with regard to the outcome of local dynamical experiments. We conclude that Newton's theory of gravity is based on a false distinction, between gravitational forces and forces induced by acceleration, and hence on an unverifiable assumption, that one can establish an inertial frame, at rest or in uniform motion, in the presence of gravitating matter.

5.6. Newtonian Gravity Reformulated

Einstein noticed that by considering the Principle of Equivalence from another point of view one could reconstruct a consistent theory of gravity. Instead of using the Equivalence Principle to create gravity in an accelerated frame, one could make a suitable choice of accelerated frame to remove the effects of gravity in local observations.

Consider an observer subject to no non-gravitational forces. We say that such an observer is in *free-fall*, or *freely falling*. Astronauts in Earth orbits floating freely in their spacecraft are in free-fall. Whatever gravitational field appears to be present to an Earth-bound observer, the astronaut will see that objects in his local environment satisfy the laws of Newtonian dynamics exactly (apart from relativistic corrections, of course). Subjected to no non-gravitational forces they will experience no acceleration. This result depends on the Equivalence Principle: for an Earth-bound observer all free objects in the spacecraft experience the same accelerations which become, from the point of view of the astronaut, the same zero accelerations. The conclusion would not hold for, say, the electromagnetic force for which the accelerations depend on the charges of the bodies and not just on the electromagnetic field.

We suppose that we know the laws of dynamics in the absence of gravity. Consequently we know how to predict the outcome of dynamical experiments in freely falling frames in gravitational fields without the need for a theory of gravity. In this sense the freely falling frames here play the same role that inertial frames of reference play in the absence of gravity. To take account of the local effects of gravity, therefore, we simply transcribe the predictions from the frame in free-fall to the frame for which they are required. We do not have to identify a mythical state of rest or uniform motion by imagining what would happen if gravity were to be switched off. We do have to identify the trajectories of freely falling bodies; but this is observationally straightforward: we simply observe the trajectories of bodies subject to no non-gravitational forces in some convenient reference frame. The description of the falling apple becomes for Einstein not primarily a question of the gravitational force that causes the apple to fall, but a question of the non-gravitational force required to prevent the apple from falling freely while it is still attached to the tree.

To describe the free-falls of Newtonian theory conveniently, we choose a system of coordinates (x, y, z) such that those test bodies which are sufficiently far removed from all matter that their environment may be considered gravity-free, move with constant velocities in these coordinates, i.e. far from matter these are standard inertial coordinates. The free-falls are then given as

$$\frac{d^2 \mathbf{x}}{dt^2} + \nabla\phi = 0, \qquad (5.6.1)$$

where $\mathbf{x} = (x, y, z)$ and ϕ is the gravitational potential. Note that we transform this to the form of Newton's second law by multiplying (5.6.1) by the inertial mass of a test body. The passive gravitational mass never appears in this new formalism.

The task of a theory of gravity is to predict these free-falls from the distribution of gravitating matter, or, equivalently, to provide a value for ϕ. In Newtonian theory this is given by the Poisson equation (5.3.1). Of course, all we have achieved is an alternative *description* of Newtonian theory; the results are essentially unchanged (with the rather academic possibility of some slight generalisation which appears in a more thorough analysis). The advantage of this apparently trivial achievement will be evident when we come to see how relativistic mechanics must be modified in the presence of gravity. For there we do not have a theory at all to start with, and without this alternative viewpoint we should not be able to obtain one.

5.7. Gravity and the Deviation of Free-falls

It is important to realise that the Principle of Equivalence does not imply that all the effects of gravity can be removed by passing to a single accelerated reference frame — if it did the theory of gravity would be entirely trivial. We have been careful to refer always to *local* effects, by which we mean effects which do not depend on the variation of the acceleration due to gravity over the system under discussion. Only these effects can be eliminated by choice of frame. For a large system on the Earth the acceleration due to gravity will vary in direction over the system, since the acceleration will always be towards the centre of the Earth. There is therefore a tendency for widely separated particles falling freely towards the centre of the Earth to accelerate towards each other. This has nothing to do with their mutual gravity which can be made negligible.

These non-local effects are called *tidal* forces since the tides on Earth are an example of the differential acceleration of gravity — due in this case to the variation of the Moon's gravitational attraction over the diameter of the Earth. By definition, a local experiment is therefore one in which the region of space occupied, and the duration of observation, are sufficiently small that tidal effects do not appear to the level of accuracy considered. Experiments over larger regions or of greater accuracy will reveal the convergence or divergence of freely falling particles. Only in the complete absence of matter will tidal forces be absent and will freely falling bodies move with constant relative velocity. Conversely, the measurement of tidal forces and deviating free-falls reveals the presence of gravitating matter.

It is important to realise also that this does not compromise our discussion of the Equivalence Principle. A general dynamical system can be considered to be composed of local subsystems which can be analysed in terms of the Equivalence Principle. Putting the parts together one obtains the behaviour of the whole system. It is only in actually carrying out this synthesis, by setting up and solving appropriate differential equations, that the technical complications of this approach arise. In principle, the theory of gravity comprises the prediction of freely falling trajectories and the analysis of systems in free-fall. The dynamics of the Universe translates into the geometry of free-falls.

5.8. The Principle of Equivalence in Relativistic Physics

The Galilean Principle of Equivalence refers to accelerations in space measured in terms of Newtonian absolute time. In relativistic physics we know there is no privileged way of slicing up space–time into spaces of absolute simultaneity. We therefore reformulate the Principle of Equivalence in a space–time context:

> *The 'weak' form of the Principle of Equivalence* states that in the absence of non-gravitational influences, test bodies released from the same point at the same time with the same initial velocity will follow identical trajectories in space–time independently of their composition and internal structure.

To be strictly correct we should refer to test bodies at neighbouring points initially, since two bodies cannot be at exactly the same point at one time. For obvious reasons an alternative name for this Principle is the *Universality of Free-fall.* Since it is simply a relativistic reformulation of the Galilean Equivalence Principle, it is supported by the same empirical evidence to the same high accuracy. Similarly, one can argue exactly as for the non-relativistic theory to conclude that it is impossible by means of local dynamical experiments to distinguish between gravitational forces and forces due to acceleration, even if we use bodies for which relativistic rather than Newtonian dynamics is appropriate. This conclusion is quite independent of the as yet unknown relativistic form for the gravitational force, and depends only on the assertion that the acceleration is the same for all bodies.

We should like to proceed by analogy with Newtonian theory. There we reconstructed the theory of gravity by taking Newtonian dynamics to hold in its standard form in a freely falling frame. Here we should like to obtain a relativistic theory of gravity by taking relativistic mechanics to be valid locally in a freely falling frame, but unfortunately we cannot quite do this. For while Newtonian dynamics is the whole of Newtonian physics, relativistic dynamics is only a part of relativistic physics. It is therefore possible to argue that although we cannot distinguish gravity and acceleration by dynamical experiments, we might be able to perform other types of experiment, with light for example, which would provide a distinction. In order to rule out such a possibility we need a stronger form of the Principle of Equivalence referring to all physical experiments, not just dynamics:

> *The 'strong' form of the Principle of Equivalence* states that in a freely falling frame the laws of non-gravitational physics assume the standard form they have in the absence of gravity.

This is sometimes called the *Einstein Principle of Equivalence.* For our purposes the standard form of the laws of physics means their special relativistic form. In view of the fact that we do not yet know *all* the laws of physics this form of the Principle is somewhat difficult to verify by experiment, although attempts have been made to deduce it from the weak form. The idea is that a violation of special relativistic physics in a freely falling frame will affect the internal composition of a test body in such a way as to cause it to deviate from universal free-fall, and hence to violate the weak principle. It appears that this deduction is possible provided one makes certain assumptions concerning the form of physical laws and rules out certain possible accidental cancellations of effects (Lightman and Lee 1973). We shall assume that the strong form of the Equivalence Principle is valid. This is a crucial assumption since we shall see that an appropriate interpretation of this Principle commits us to a description of gravity in terms of the geometry of space–time (Will 1979, Thorne *et al* 1973).

5.9. Space–Time Geometry in Special Relativity

A key feature of special relativity is the existence of a useful measure of the separation of events in space–time. Suppose the event Q in space–time lies in the future light cone of an event P; Q can then be reached by a unique inertial observer passing through P. The time recorded by a clock carried by the observer between P and Q is a measure of the space–time interval τ_{PQ} between the points. The interval recorded depends not only on the events P and Q, but also on the path between them; consequently we specify an inertial trajectory in the definition. In Newtonian theory this construction is not very useful since it amounts merely to measuring the absolute time difference between P and Q. This is independent of their spatial separation and of the trajectory that joins them.

If P and Q cannot be joined by an inertial path (with velocity less than that of light) then there exists an inertial observer for whom P and Q are simultaneous events. We can use the spatial distance between them s_{PQ} as determined by this observer as a measure of the interval and set $\tau_{PQ}^2 = -s_{PQ}^2$. Recall that the minus sign is required since $c^2 t_{PQ}^2 - |x_{PQ}|^2$ is the general form for the square of an invariant interval, i.e. one which is the same for all observers in uniform relative motion. The geometry of space–time means just the properties of the function τ_{PQ} for all pairs of points.

It is more convenient to work with a function of neighbouring points from which τ_{PQ} may be constructed rather than carry the directly measurable information on the interval between all pairs of points. Let P and Q be neighbouring points with coordinates $(x^\mu) = (x^0, x^1, x^2, x^3) \equiv (ct, x^1, x^2, x^3)$ and $(x^\mu + dx^\mu)$, respectively, in a Minkowski coordinate system. The use of superscript indices is explained in §5.14, but may be simply regarded as a convention here. The square of the interval between P and Q is given by

$$- d\tau_{PQ}^2 = -(dx^0)^2 + (dx^1)^2 + (dx^2)^2 + (dx^3)^2. \tag{5.9.1}$$

From this we can construct the interval between finitely separated events along any path, not merely an inertial one. In principle, we add together segments of straight lines joining nearby points, and in practice this is achieved by integration of $d\tau$ along the path. In particular we can reconstruct τ_{PQ} to find

$$- \tau_{PQ}^2 = -(x_P^0 - x_Q^0)^2 + (x_P^1 - x_Q^1)^2 + (x_P^2 - x_Q^2)^2 + (x_P^3 - x_Q^3)^2 \tag{5.9.2}$$

in an obvious notation.

The geometry of the Minkowski space-time of special relativity is contained in (5.9.2) or equivalently in (5.9.1). Consequently one cannot derive these expressions (except from each other); they are postulated as the basis of the theory. Furthermore, they serve to define what we mean by Minkowski coordinates: if the coordinate differences can be combined as in (5.9.2) to yield the measured interval, then they are Minkowski coordinates.

For conciseness of presentation we introduce the matrix $(\eta_{\mu\nu})$ by

$$(\eta_{\mu\nu}) = \begin{pmatrix} -1 & 0 & 0 & 0 \\ 0 & 1 & 0 & 0 \\ 0 & 0 & 1 & 0 \\ 0 & 0 & 0 & 1 \end{pmatrix} \tag{5.9.3}$$

and write

$$-d\tau^2 = \sum_{\nu=0}^{3} \sum_{\mu=0}^{3} \eta_{\mu\nu} \, dx^\mu \, dx^\nu$$

in place of (5.9.1). Furthermore, we can use the Einstein summation convention, whereby any index that occurs twice is understood to be summed over from 0 to 3, and we drop the explicit summation sign. Thus

$$-d\tau^2 = \eta_{\mu\nu} \, dx^\mu \, dx^\nu \tag{5.9.4}$$

is understood as a shorthand form of (5.9.1). This expression is called the Minkowski metric and the $\eta_{\mu\nu}$ are the Minkowski metric coefficients.

The Minkowski metric determines the relations between intervals in space-time analogously to the way in which Pythagoras' theorem determines relations between distances in Euclidean space. In addition, it determines the geometry of free-falls and of light rays. Free-falls, or inertial trajectories, are determined by the condition that they be straight lines in Minkowski coordinates. Hence they must satisfy

$$\frac{d^2 x^\mu}{d\tau^2} = 0, \tag{5.9.5}$$

which are simply the differential equations of a straight line in space-time parametrised by τ, and expressed in coordinates in which the metric has the form (5.9.4). This is the first law of relativistic dynamics. A light ray is determined by the additional condition that $d\tau^2 = 0$ for any two neighbouring points on the ray. This can be shown to follow from Maxwell's equations, which provide the relativistic theory of light, or directly from the invariance of the velocity of light.

5.10. The Geometry of Space–Time

According to the Principle of Equivalence, a clock of sufficiently small dimensions falling freely in a gravitational field behaves as it would in the absence of gravity.

Therefore we can construct a measure of the interval τ_{PQ} between any two events P and Q which can be joined by a free-fall. The properties of the function τ_{PQ} are the geometry of space-time in the presence of gravity. If τ_{PQ} could be given the form (5.9.2) by suitable choice of coordinates then the free falls would satisfy (5.9.5); two initially parallel free-falls would remain parallel, so there would be no tidal effects, and hence no gravity. Therefore, in the presence of gravitating matter, τ_{PQ} cannot be given the form (5.9.2).

On the surface of the Earth the distances between widely separated points do not satisfy Pythagoras' theorem; we describe this by saying that the surface is *curved*. By analogy, if the free-fall time relations between widely separated events in space-time do not satisfy (5.9.2) we say that the *space-time* is curved. The curvature of the Universe refers to nothing more than the timetable for the free-falls between galaxies. Notice in particular that this is expressed entirely in terms of measurements within the Universe. Not only is one not required to imagine four-dimensional space-time as a curved surface, but it is actually inappropriate to do so. In an expanding Universe, the free-fall time taken to return home from a distant galaxy will be longer than the time taken to reach it, essentially by virtue of the meaning of expansion. In this way, the dynamics of the Universe translates into the geometry of space-time.

It is more convenient to describe the geometry in terms of the time between neighbouring points, since this turns out to involve the tabulation of ten functions of the coordinates of a single point, rather than a function of two points. Since we do not know the results of measurements until we have set up the geometry, we cannot begin by constructing coordinates in terms of distances and times. So we simply assume that some continuous labelling scheme for events has been chosen, the significance of which will become apparent once we know the geometry.

Letting the points P and Q have arbitrarily chosen labels (x^λ) and $(x^\lambda + dx^\lambda)$, respectively, we write $\tau_{PQ}^2 \equiv \sigma_P(x^\lambda + dx^\lambda)$ merely to simplify the notation. Then, by Taylor's theorem,

$$\sigma_P(x^\lambda + dx^\lambda) = \sigma_P(x^\lambda) + \left[\frac{\partial \sigma_P(y^\lambda)}{\partial y^\mu}\right]_{y^\lambda = x^\lambda} dx^\mu + \frac{1}{2}\left[\frac{\partial^2 \sigma_P(y^\lambda)}{\partial y^\mu \partial y^\nu}\right]_{y^\lambda = x^\lambda} dx^\mu\, dx^\nu + \dots$$

$$(5.10.1)$$

We write the derivatives in this way to make it clear that the point P is kept fixed. Now $\sigma_P(x^\lambda) = 0$ since it is just the square of the time between P and itself. Similarly, the second term on the right-hand side of (5.10.1) is the first derivative of the square of the time interval, which is proportional to the time interval itself, and so vanishes when evaluated at P. Using the conventional concise notation, we write the surviving term as

$$-d\tau^2 = g_{\mu\nu}(x^\lambda)\, dx^\mu\, dx^\nu, \qquad (5.10.2)$$

with $g_{\mu\nu} = g_{\nu\mu}$. The higher-order terms are, of course, to be neglected in the limit $dx^\mu \to 0$.

The expression (5.10.2) defines the metric coefficients $g_{\mu\nu}(x^\lambda)$ as functions of the arbitrary labels (x^λ). In addition, it gives the meaning of those coordinate labels since it associates a physically measured time (or distance) with pairs of points separated by differences in coordinates of amount (dx^λ). The numerical value of these coordinate differences will depend on the choice of coordinates, as will the numerical values of the metric coefficients; the results for the physical measurements of distance and time will not. The coordinates themselves do not need to have any direct physical meaning; they acquire it through the expression for the metric. Of course, if a situation has special symmetry it may be possible to choose coordinates with a direct meaning and such a choice may considerably simplify calculations. For example, in the Minkowski space–time of special relativity it is often advantageous to choose Minkowski coordinates (5.9.4) which are directly related to measured times and distances. When one is dealing with the gravitational field of a general distribution of matter, and such symmetry is absent, it is usually impossible to choose coordinates with direct physical significance. In fact, in such cases, the meaning of the coordinates is only provided after one has solved the problem of finding what the metric coefficients are, a problem which we consider in §5.13.

Just as is the case in special relativity, the metric (5.10.2) provides us with a way of associating an interval between any pair of points measured along any path. In principle we add up the interval along small straight line segments of the path; in practice this means we integrate $d\tau$ along the path. In particular, we can obtain a unique interval between pairs of points which cannot be joined by a free-fall. From all the paths which join the points we pick that one which extremises the magnitude of the interval along the path obtained by integrating $d\tau$, and use this extremal value as our measure of the separation of the points. This is always possible for points which are not too widely separated.

There are ten distinct metric coefficients, since from the definition (5.10.1) we have $g_{\mu\nu} = g_{\nu\mu}$. Now we can always arrange that the Minkowski metric coefficients in (5.9.4) be replaced by certain functions of position by a suitably obtuse choice of coordinates. As a trivial example, instead of the Minkowski coordinates (x^0, x^1, x^2, x^3) we might choose polar coordinates (ct, r, θ, ϕ), which are defined in the usual way by

$$x^0 = ct \qquad\qquad x^1 = r \sin\theta \, \cos\phi$$
$$x^2 = r \sin\theta \, \sin\phi \qquad\qquad x^3 = r \cos\theta$$

(5.10.3)

We obtain

$$-d\tau^2 = -c^2 \, dt^2 + dr^2 + r^2 \, d\theta^2 + r^2 \sin^2\theta \, d\phi^2$$

by straightforward substitution of (5.10.3) in (5.9.4). However, there are only four coordinates to be transformed and ten metric coefficients. Therefore we cannot in general transform the Minkowski metric to the form (5.10.2) with arbitrary $g_{\mu\nu}$. Conversely, of course, we cannot in general transform the expression (5.10.2) to the Minkowski form. If it so happens that we can perform such a transformation, then the free-falls have the form (5.9.5), there are no tidal forces and hence no gravitating bodies. The curvature of space–time is therefore expressed through the generality of the metric coefficients in (5.10.2).

Suppose that at each point of space–time we know the acceleration of a local freely falling observer relative to some convenient but arbitrary reference system. This information is sufficient to enable us to construct the freely falling trajectories. For, starting at some initial point with some initial velocity we construct a segment of a path with the given acceleration of free-fall at the initial point. After some time we reach an event where the free-fall acceleration is significantly different from the initial one. So we proceed with this new acceleration, and so on. In this way we approximate to a smooth curve which is the desired freely falling trajectory.

The important point now is that the information required to start with, the accelerations of local freely falling observers, is contained in the metric coefficients $g_{\mu\nu}(x^\lambda)$. The convenient but arbitrary reference system is the coordinate system (x^λ). In a neighbourhood of each point of space–time a freely falling observer must find that he can construct a system of coordinates in which the metric takes the Minkowski form to a good approximation. If we can find a mathematical transformation of coordinate labels which takes us from the known $g_{\mu\nu}(x^\lambda)$ to the required $\eta_{\mu\nu}$ with sufficient accuracy then we know how to obtain the labelling used by the freely falling observer. We shall see how this purely mathematical problem can be solved in a particular example in the next section. From a knowledge of the transformation one can clearly work out the acceleration of one system relative to the other, and this is the information we require. We conclude that the metric coefficients determine the free-fall trajectories. A light ray is determined by the additional condition that $d\tau^2 = 0$ for any two neighbouring points on the ray. This follows from the strong form of the Principle of Equivalence since it is a local condition. It follows that the metric coefficients provide all the information we require for the solution of problems in particle dynamics in the presence of gravity. Indeed, when proper care is taken of tidal forces it follows from the above argument (and the Principle of Equivalence) that the behaviour of all physical systems in the presence of gravity is determined once we know the metric coefficients throughout space–time.

5.11. Geodesics

As a matter of terminology the free-fall trajectories in space–time are called *geodesics*. This relates to the fact that the differential equation defining them is formally identical to the defining equation of the curves of shortest length between pairs of points on a curved surface. We shall not make use of this result, but we shall instead obtain the geodesic equation from the physical considerations we have already given, namely from the fact that the geodesic equation must reduce to (5.9.5) in a coordinate system which is locally approximately Minkowskian. In order not to confuse the derivation in clouds of suffixes and superscripts, we shall consider here only a particular example; that is, we choose a particular form for the metric coefficients which will be of relevance later. The calculation then reduces to an exercise in partial differentiation. The completely general result may be obtained by exactly the same method.

Consider the metric given by

$$- d\tau^2 = - dt^2 + [a(t)]^2 \, [(dx^1)^2 + (dx^2)^2 + (dx^3)^2].$$
(5.11.1)

Recall that the metric itself specifies the meaning of the coordinates (t, x^1, y^2, z^3) and we shall extract this in due course. Note here though that there is no factor of c^2 multiplying dt^2, so t cannot quite be a time in SI units. Relativists conventionally put $c = 1$ in calculations in order to save writing and to avoid prejudging the physical meaning of the coordinates. This practice is equivalent to using light-seconds to measure distance. The symbol t becomes time in seconds if we replace t in the following equations by t/c. The function $a(t)$ is arbitrary; in applications we shall find that the form $a(t) \propto t^\alpha$ (with $\alpha = \frac{1}{2}$ or $\frac{2}{3}$) is of some considerable importance.

We want to find new coordinates – call them $(\lambda, \xi^1, \xi^2, \xi^3)$ say – in which the metric approximates the form (5.9.4) in the neighbourhood of a fixed, but arbitrary point P with coordinates $(t_P, x^1_P, y^2_P, z^3_P)$ in the original system. First we must decide exactly what degree of accuracy we require of the approximation. We shall save a lot of writing by using a vector notation, such as (λ, ξ), for the coordinate labels. Then let $g_{\mu\nu}(\lambda, \xi)$ be a typical metric component in the new coordinates. Expand it about $P \equiv (\lambda_P, \xi_P)$ by Taylor's theorem to get

$$g_{\mu\nu}(\lambda, \xi) = g_{\mu\nu}(\lambda_P, \xi_P) + \left\{ (\lambda - \lambda_P) \left[\frac{\partial g_{\mu\nu}}{\partial \lambda} \right]_P + (\xi^i - \xi^i_P) \left[\frac{\partial g_{\mu\nu}}{\partial \xi^i} \right]_P \right\}$$

$$+ \left\{ \tfrac{1}{2}(\lambda - \lambda_P)^2 \left[\frac{\partial^2 g_{\mu\nu}}{\partial \lambda^2} \right]_P + (\lambda - \lambda_P)(\xi^i - \xi^i_P) \left[\frac{\partial^2 g_{\mu\nu}}{\partial \xi^i \, \partial \lambda} \right]_P \right.$$

$$\left. + \tfrac{1}{2}(\xi^j - \xi^j_P)(\xi^i - \xi^i_P) \left[\frac{\partial^2 g_{\mu\nu}}{\partial \xi^i \, \partial \xi^j} \right]_P \right\} + \dots$$
(5.11.2)

using the summation convention on repeated indices.

We require that the first term be $\eta_{\mu\nu}$ so that (5.9.4) will be exact at P. Since $[g_{\mu\nu}(\lambda_P, \xi_P)] = (g_{\mu\nu})_P$ is just a constant symmetric matrix, we can reduce it to the diagonal form (5.9.3) by a suitable transformation of axes in the standard way. If this is not possible, then the Equivalence Principle is violated, and the supposed metric under consideration is not an acceptable metric for space–time.

Any acceptable metric can therefore be written in the form (5.11.2), with $\eta_{\mu\nu}$ as the first term, with only a trivial restriction on the coordinates at P. (Essentially they have to be orthogonal coordinates at P.) A better approximation is obtained the more terms of the series (5.11.2) can be made to vanish. In particular, we attempt to define (λ, ξ) such that the linear terms in the expansion are zero. The metric will then approximate the Minkowski form to second order in the displacement from P. This is the best that can be done; one cannot in general demand that all the second-order terms of the expansion vanish.

The most general transformation between the coordinates (t, x) and (λ, ξ) in the neighbourhood of P can itself be written as a Taylor expansion. For example,

$$(t - t_P) = \Lambda_0(\lambda - \lambda_P) + \Lambda_i(\xi^i - \xi_P^i) + \tfrac{1}{2}\Lambda_{00}(\lambda - \lambda_P)^2 + \Lambda_{i0}(\lambda - \lambda_P)(\xi^i - \xi_P^i)$$
$$+ \tfrac{1}{2}\Lambda_{ij}(\xi^i - \xi_P^i)(\xi^j - \xi_P^j) + \dots$$

using, as usual, the summation convention for repeated indices, with a similar expression for $(x^i - x_P^i)$. The Λ are all constants and the object of the exercise is to find their values. In principle one proceeds to compute dt and dx in terms of $d\lambda$ and $d\xi$, to substitute in (5.11.1) and impose the condition that the terms linear in $(\lambda - \lambda_P)$ and $(\xi - \xi_P)$ vanish. In practice, one can simplify the coordinate transformation by guessing that some of the terms, like Λ_i for example, will be zero. We shall simplify the computation even further by presenting the result and verifying that it is correct. We assert that

$$t - t_P = \lambda - \lambda_P - \tfrac{1}{2}(\dot{a}/a)_P(\xi - \xi_P)^2$$
$$x - x_P = (1/a)_P(\xi - \xi_P) - (\dot{a}/a^2)_P(\lambda - \lambda_P)(\xi - \xi_P) \qquad (5.11.3)$$

where $\dot{a} \equiv da/dt$, is the required transformation. Thus we have

$$dt = d\lambda - (\dot{a}/a)_P(\xi - \xi_P) \cdot d\xi$$
$$dx = (1/a)_P d\xi - (\dot{a}/a^2)_P(\lambda - \lambda_P) d\xi - (\dot{a}/a^2)_P(\xi - \xi_P) d\lambda. \qquad (5.11.4)$$

To first order, equation (5.11.1) becomes

$$-d\tau^2 = -dt^2 + (a^2)_P[1 + 2(\dot{a}/a)_P(t - t_P)][dx^2].$$

Substituting (5.11.4) and retaining only first-order terms yields

$$-d\tau^2 = -\{1 + 0[(\xi - \xi_P)^2]\} d\lambda^2 + \{1 + 0[(\xi - \xi_P)^2] + 0[(\lambda - \lambda_P)^2]\} d\xi^2,$$

which is the required result.

We can now readily obtain the equations determining the geodesics. In (λ, ξ) coordinates at P these are exactly

$$\frac{d^2\lambda}{d\tau^2} = 0; \qquad \frac{d^2\xi}{d\tau^2} = 0. \qquad (5.11.5)$$

To transform to (t, x) coordinates we first solve (5.11.3) or (5.11.4) for (λ, ξ) as functions of (t, x). Thus, to first order,

$$d\lambda = dt + (a\dot{a})_P(x - x_P) \cdot dx$$
$$d\xi = a_P dx + \dot{a}_P(t - t_P) dx + \dot{a}_P(x - x_P) dt.$$

Equations (5.11.5) therefore become

$$t'' + (a\dot{a})_P (x')^2 + (a\dot{a})_P (x - x_P) \cdot x'' = 0$$

$$x'' + 2(\dot{a}/a)_P \, t'x' + (\dot{a}/a)_P (t - t_P) x'' + (\dot{a}/a)_P (x - x_P) \, t'' = 0$$

where the primes indicate differentiation with respect to τ. At the point P, these simplify *exactly* to

$$t'' + a\dot{a}(x')^2 = 0$$

$$x'' + 2\dot{a}/a \, t'x' = 0. \qquad (5.11.6)$$

But P is an arbitrary point. Therefore (5.11.6) are *exactly* the required differential equations of the geodesics.

A solution of (5.11.6), consistent with (5.11.1), is

$$t = \tau; \qquad x = \text{constant}.$$

Therefore particles at constant values of x are in free-fall and t is the time measured by clocks falling with such particles. Here then the coordinate t has a direct significance. In successive surfaces, given by $t = $ constant, the square of the distance $ds^2 = - d\tau^2$ between these particles in free-fall is proportional to $[a(t)]^2$. *In this sense* these freely falling particles are moving away from (or towards) one another. We shall see later that this metric represents an expanding universe, the role of these freely falling particles being played by clusters of galaxies.

5.12. Curvature and Geodesic Deviation

Return for a moment to the Newtonian theory of gravity. The free-falls are given by (5.6.1)

$$\frac{d^2 x^i}{dt^2} + \frac{\partial \phi}{\partial x^i} = 0 \qquad (i = 1, 2, 3)$$

which is the analogue of the geodesic equation. If a neighbouring geodesic is labelled by $x^i(t) + \epsilon^i(t)$, for small $\epsilon^i(t)$, then

$$0 = \frac{d^2}{dt^2} (x^i + \epsilon^i) + \left[\frac{\partial \phi}{\partial x^i}\right]_{x+\epsilon} \approx \frac{d^2 x^i}{dt^2} + \frac{d^2 \epsilon^i}{dt^2} + \left[\frac{\partial \phi}{\partial x^i}\right]_x + \left[\frac{\partial^2 \phi}{\partial x^i \partial x^j}\right]_x \epsilon^j.$$

Subtracting, we obtain

$$\frac{d^2 \epsilon^i}{dt^2} + \frac{\partial^2 \phi}{\partial x^i \partial x^j} \epsilon^j = 0. \qquad (5.12.1)$$

This equation gives the deviation of free-falls; tidal forces exist if the second derivatives of the potential are not all zero.

We now want to derive a similar equation of geodesic deviation for the metric (5.11.1). This will enable us to illustrate how the curvature of space–time appears as

102

second derivatives of the metric coefficients governing the deviation of geodesics. If all of these second derivatives were zero everywhere the geodesics would not deviate and tidal forces would be absent. Therefore we demonstrate by example that the second-order terms in the expansion (5.11.2) cannot all be made to vanish throughout space–time.

In the previous section we set up a system of coordinates which approximated Minkowski coordinates in the neighbourhood of a point. It is possible (and for our purposes here, rather more convenient) to do more than this and construct a system of coordinates which approximate Minkowski coordinates along a single geodesic. The meaning of the approximate agreement of coordinates here is the same as in §5.11; we require the metric to depart from its Minkowski form in the neighbourhood of the fiducial geodesic only in terms of the second order and higher in the distance from the curve. Clearly we are representing here the local Minkowski coordinates appropriate to the freely falling observer throughout his history and not just in the neighbourhood of a single event.

In the space–time described by the metric (5.11.1) let the fiducial geodesic be chosen as $\mathbf{x} = 0$ for convenience. We seek new coordinates (λ, ξ) related to $(t, \mathbf{x})$ by a transformation of the form

$$t = t_0(\lambda) + t_2(\lambda)\, \xi^2 \tag{5.12.2}$$

$$\mathbf{x} = x_1(\lambda)\, \xi + x_2(\lambda)\, \xi^2 \tag{5.12.3}$$

such that the metric expressed in (λ, ξ) coordinates has the Minkowski form, $-d\lambda^2 + d\xi^2$, to first order in ξ for arbitrary λ. Clearly, a certain amount of guesswork has gone into restricting the form of the coordinate transformation. In particular there is no linear term in ξ in (5.12.2) since the rotational invariance of the spatial part of (5.11.1) means we do not expect the transformation to depend on spatial direction. The term independent of ξ is absent from (5.12.3) since we want $\xi = 0$ on the fiducial geodesic.

Substitution of the differential forms of (5.12.2) and (5.12.3) into the metric (5.11.1), and comparison with the Minkowski form to first order in ξ, leads to the following result:

$$t = \lambda - \tfrac{1}{2}\xi^2 a'(\lambda)/a(\lambda)$$
$$\mathbf{x} = \xi/a(\lambda), \tag{5.12.4}$$

where primes indicate differentiation with respect to λ. It is straightforward to check this by direct substitution. Retaining second-order terms, we obtain

$$d\tau^2 \approx -d\lambda^2 \left(1 - \xi^2\, \frac{a''(\lambda)}{a(\lambda)}\right) + d\xi^2 \left[1 - \xi^2 \left(\frac{a'(\lambda)}{a(\lambda)}\right)^2\right] - \left(\frac{a'(\lambda)}{a(\lambda)}\right)^2 (\xi \cdot d\xi)^2, \tag{5.12.5}$$

correct to second order. Neglecting higher-order terms, the deviation of geodesics is given by the second-order deviation from the Minkowski form in (5.12.5).

103

We want to find the equation of a geodesic near $\xi = 0$ in the metric (5.12.5). In principle, we simply repeat the procedure of §5.11, picking a point $P(\lambda_P, \xi_P)$, and finding a coordinate transformation which reduces the metric (5.12.5) to Minkowski form exactly at P, and approximately in a neighbourhood of P. In practice, of course, we already know the equations of arbitrary geodesics in (t, x) coordinates from §5.11, so we can simply transform these to the coordinates (λ, ξ) of equation (5.12.4). Retaining only first-order terms, we obtain, along the fiducial geodesic,

$$\ddot{\lambda} = 0 \tag{5.12.6}$$

$$\ddot{\xi} = (a''/a)\,\dot{\lambda}^2 \xi + 2(a'/a)^2(\xi \cdot \dot{\xi})\,\dot{\xi} \tag{5.12.7}$$

where the dots indicate differentiation with respect to τ, and the primes differentiation with respect to λ. From equation (5.12.6) we see that, apart from a possible trivial rescaling and change of origin, we have $\tau = \lambda$, and from (5.12.4) it follows that t is proper time along the fiducial geodesic. Since ξ measures the displacement of a point on the geodesic from the fiducial geodesic, equation (5.12.7) is the precise analogue of the Newtonian equation (5.12.1).

For the deviation of geodesics given by $x = $ constant, we have $\xi = a(\lambda) x$ from (5.12.4), and hence $\dot{\xi} = a'\dot{\lambda} x = (a'/a)\,\dot{\lambda}\xi$. If we take $\tau = \lambda = t$, we obtain the important result that these neighbouring geodesics have a relative velocity $(\dot{a}/a)\,\xi$. Furthermore, for these geodesics the final term of equation (5.12.7) becomes $-2(\dot{a}/a)^4\xi^2\xi$. The fact that this is not zero does not imply that we have made a mistake! For this surviving term is of third order in $|\xi|$ and so is to be neglected to the order of approximation under consideration. This is an important point for the following reason. Clearly, we could contemplate the retention of higher-order terms in the metric (5.12.5). The values of these terms could be adjusted by the addition of higher-order terms to the coordinate transformation (5.12.4). The fact that the geodesic deviation equation (5.12.7) is not satisfied at the third order along exact geodesics suggests that the addition of third-order terms to (5.12.4) could actually alter the form of (5.12.7); and this is indeed the case. It means that not all of the terms in this equation have a direct physical significance independent of the choice of coordinates, a point which will be relevant in §5.13, and will be discussed more fully in §5.14.

5.13. Curvature and Matter

So far we have seen how the effects of gravity must be described in terms of the geometry of space–time, and how we can therefore, at least in principle, compute the behaviour of physical systems in a known gravitational field by passing to a freely falling frame. We have not yet touched on the equally important other half of the problem, namely, how we can find the space–time geometry appropriate to a given distribution of matter. In Newtonian theory the analogous problem of finding the gravitational potential is solved by postulating Poisson's equation (5.3.1), which is equivalent to the inverse square law of force generalised to an arbitrary matter distribution. In general relativity we therefore attempt to construct an analogous set of equations to determine the metric coefficients.

The Newtonian equation of geodesic deviation (5.12.1) shows that the second derivatives of the potential, $\partial^2\phi/\partial x^i\,\partial x^j$, control the deviation, while the sum of such terms, $\nabla^2\phi$ appears in Poisson's equation. Our first attempt to guess some suitable field equations for the metric, therefore, might be to perform an exactly analogous step in relativity. We shall carry out the argument for the particular example of a metric of the form (5.11.1). This simplifies the algebra considerably, although of course we lose all generality. Nevertheless, the form of the argument is not restricted to just this particular example.

We are seeking an equation for $a(t)$. The first guess is simply to add coefficients in the geodesics deviation equation (5.12.6). However, this cannot be quite right because some of these terms can be changed by a different choice of coordinates. Let us agree to fix $\lambda = \tau = t$, since this coordinate then has the physical meaning of proper time measured by a freely falling observer. It then turns out that the inclusion of higher-order terms in ξ in the second of equations (5.12.4) cannot alter the first term on the right-hand side of the deviation equation (5.12.7) (see §§5.14 and 6.6). A second guess is therefore to add just the coefficients in this term to obtain the analogue of Poisson's equation. This would give

$$-3\frac{\ddot{a}}{a} = \frac{4\pi G\rho}{c^2} \tag{5.13.1}$$

as the equation determining $a(t)$ from a known $\rho(t)$. We should not expect this to be a satisfactory guess if the particles in the Universe have large random motions, for in this case the velocity terms involving $\dot{\xi}$ in (5.12.7) might make some significant contribution.

In Newtonian theory $\rho(t)$ is quite arbitrary, at least in principle. This is so because there is no equivalence of mass and energy built into Newtonian physics, and hence there is the possibility of the arbitrary interconversion of gravitating matter and non-gravitating energy. Of course, this is not necessarily how one uses Newtonian theory in practice, since mass usually is conserved. Nevertheless, it explains why in Newtonian gravity one need go no further than Poisson's equation, whereas in relativity we cannot stop at (5.13.1), because we must take into account the equivalence and conservation of mass and energy.

Suppose then that the Universe consists of point particles with no random motion, equivalent to a gas at zero temperature and pressure. This is actually quite a good approximation in applications in which the particles can be taken to be clusters of galaxies. This system has no internal energy other than its rest mass, so the total rest mass of a given set of particles must be conserved. This is just an expression of the conservation of galaxies. Therefore

$$\rho(t)\,\Delta\xi^1\,\Delta\xi^2\,\Delta\xi^3 = \text{constant}$$

for the volume $\Delta \xi^1 \Delta \xi^2 \Delta \xi^3$, or

$$\rho(t) a^3(t) \Delta x^1 \Delta x^2 \Delta x^3 = \text{constant}.$$

This latter expression is useful since we know $\Delta x_i = \text{constant}$ between a given pair of particles; so we must have

$$\rho(t) a^3(t) = \text{constant} = \rho_0 a_0^3. \tag{5.13.2}$$

Equation (5.13.2) can be substituted in (5.13.1) to eliminate ρ, whence

$$-\ddot{a} = \frac{4\pi G}{3c^2} \frac{\rho_0 a_0^3}{a^2}. \tag{5.13.3}$$

On integration, we obtain

$$\tfrac{3}{2} \left(\frac{\dot{a}}{a} \right)^2 = \frac{4\pi G}{c^2} \rho, \tag{5.13.4}$$

where we have set the constant of integration equal to zero. (So in the absence of matter $a(t) = \text{constant}$, and we regain the space-time structure of special relativity.) In this case therefore, equations (5.13.1) and (5.13.4) are equivalent: we can take either of them as the field equation. However, when one is dealing with matter which is more general than a gas at zero temperature it turns out that (5.13.4) still holds (provided we take ρ to be the total density of mass–energy, not merely the rest mass density), whereas (5.13.1) is incorrect (see §6.6).

Random motions of the particles that make up the Universe appear macroscopically as a non-zero pressure. In this case work must be done against pressure forces to expand the region occupied by a given set of particles, and this comes from the internal energy of the particles. This internal energy now includes not only their rest mass but also the kinetic energy of *random* motion. Only random motion is relevant, not the Hubble velocity of expansion, because it is the velocity of particles relative to a freely falling observer at a point that determines the local energy density there. Let the internal energy per unit volume be $\mu(t)c^2$. Equating the change in energy to the work done yields

$$\frac{d}{dt}(\mu c^2 V) = -p \frac{dV}{dt},$$

where p is the pressure and V the volume of a given set of particles. We have $V = \Delta \xi^1 \Delta \xi^2 \Delta \xi^3 = a^3(t) \Delta x^1 \Delta x^2 \Delta x^3$ for a metric of the form (5.11.1), so

$$\frac{d}{dt}(\mu a^3) = -\frac{p}{c^2} \frac{d}{dt}(a^3),$$

or, equivalently

$$\dot{\mu} + 3\dot{a}/a(\mu + p/c^2) = 0. \tag{5.13.5}$$

Equation (5.13.5) is the general form of the conservation of energy. Since all

106

energy contributes to gravity, equation (5.13.4) must be modified to

$$\frac{3}{2}\left(\frac{\dot{a}}{a}\right)^2 = \frac{4\pi G}{c^2}\mu \tag{5.13.6}$$

in this case. This is the appropriate form for the field equation required to determine one unknown metric coefficient in a metric of the form (5.11.1).

The two equations (5.13.5) and (5.13.6) are not yet complete since they now involve three unknowns, $a(t)$, $\mu(t)$ and $p(t)$. This is physically correct: we have to add a relation which describes what the system of particles is in fact made of. A universe filled with concrete may be expected to behave differently from our own. This aspect of the physics is summarised in an *equation of state* connecting pressure and density. We shall investigate the consequences of choosing particular equations of state in Chapter 6.

We can look at the system of two equations (5.13.5 and 5.13.6) from another point of view. In special relativity, space and time should be treated on an equal footing. We have obtained our field equation by considering only the deviation of time-like geodesics. One expects that the curvature of space–time is manifest mathematically also in the deviation of spatial geodesics, even though no material particles could travel on such curves. Thus in general we expect additional field equations relating other components of the curvature to the matter distribution. Of course, for a general metric there should be ten equations to determine the ten unknown metric coefficients. But even in our particular example there should be another one equation — one only because all directions in space are equivalent here. On the right-hand side of this equation we must have the spatial analogue of energy density, which is momentum density or pressure. It is a simple matter to manipulate (5.13.6) and (5.13.5) to obtain

$$\frac{2\ddot{a}}{a} + \frac{\dot{a}^2}{a^2} = -\frac{8\pi G}{c^4}p. \tag{5.13.7}$$

This is the extra field equation in this case.

Of course, only two of these three equations are independent since one of them has just been derived from the other two! The point that we illustrate here is that, in general, the field equations one guesses for the ten metric coefficients must be compatible with the conservation of energy, and, indeed, the conservation of momentum also. The field equations cannot be deduced since they constitute a postulate of the theory. The choice can be made freely subject only to self-consistency. It turns out that the requirement of energy momentum conservation severely restricts the freedom of choice for the field equations in Einstein's theory, unless one is prepared to consider more exotic types of equation containing, for example, fourth derivatives of the metric coefficients. In our particular example the change from (5.13.1) to (5.13.4) and (5.13.6) appears rather arbitrary, since we have not provided an independent route to equation (5.13.7). Nevertheless the argument does represent to a certain extent the way in which the field equations can be obtained in general. With a lot more mathematical machinery one can derive a general form with as much conviction as is usual for a postulate of a theory. The equations one obtains are called Einstein's field equations, with which we complete the description of the geometric theory of gravity provided by the general theory of relativity. Equations (5.13.6) and (5.13.7) are simplified forms of Einstein's equations appropriate to the special form of the metric of equation (5.11.1). We shall investigate the form of the field equations for slightly more general metric expressions in Chapter 6.

5.14. The Mathematics of Relativity

For the computation of particular results in the special cases of metrics taking simple forms, the methods used in other sections of this book will suffice. However, for general arguments, or for explicit computations in more complex metrics, this approach is rather clumsy. We need to develop a theory that embodies the results of piecing together infinitesimal regions of normal coordinates while avoiding the need to make explicit transformations in each particular calculation. Such a development will be sketched here in order to illustrate the general theory and to make contact with more advanced texts, but an understanding of this section is required only for parts of §§6.4 and 6.6 of this book.

We begin with the most important ingredient of mathematical theory: notation. We need to be able to refer to points in a region of space–time, so we assign them labels, or *coordinates*. In principle, these labels could be anything; in practice, it proves sensible to choose real numbers for the purpose. The coordinates do not yet have a physical meaning; they do not tell us where the points are, any more than a library shelving number tells us where to find a book in the absence of the library shelving plan. We shall provide the analogue of the shelving system in a moment. It is convenient to assume that the coordinates have been assigned in a sufficiently smooth manner, so that as we move smoothly from event to event the associated coordinates change smoothly. By a convention, the convenience of which will be revealed presently, we write the coordinate labels as (x^0, x^1, x^2, x^3), or, more briefly, as (x^λ), with the indices raised. Taking λ to start at zero is sometimes useful in space–time, since we can try to arrange that x^0 should be distinguished as a time-like coordinate. However, note that this cannot always be guaranteed from the outset, but only after the physical meaning of the labels has been ascertained.

As discussed in §5.10, the 'meaning' of the labels is given in terms of the physical distances and times separating events as computed from the metric, or 'line-element',

$$ds^2 = g_{\mu\nu} \, [(x^\lambda)] \, dx^\mu \, dx^\nu.$$

The summation convention (§5.9) is employed here and throughout this section. In order to extract the physics of the space–time from this metric we make a transformation to the 'normal coordinates' of a freely falling observer at a point P. This point is chosen arbitrarily and then fixed throughout the subsequent discussion. Thus, we seek new coordinates, (ξ^λ), in a neighbourhood of P such that

$$g_{\mu\nu}\left[(\xi^\lambda)\right] = g_{\mu\nu}(P) + O\left[(\xi^\lambda - \xi^\lambda_P)^2\right].$$

We assert that the transformation

$$\xi^\lambda - \xi^\lambda_P = x^\lambda - x^\lambda_P + \tfrac{1}{2}\Gamma^\lambda_{\mu\nu}(P)\,(x^\mu - x^\mu_P)\,(x^\nu - x^\nu_P) \qquad (5.14.1)$$

will accomplish this for a suitably chosen set of coefficients, $\Gamma^\lambda_{\mu\nu}$, the values of which will in general depend on the choice of P. We proceed to a proof.

Note first that $\Gamma^\lambda_{\mu\nu}$ may be taken to be symmetric in the indices μ,ν; i.e. $\Gamma^\lambda_{\mu\nu} = \Gamma^\lambda_{\nu\mu}$. For suppose not. Then we can write

$$\Gamma^\lambda_{\mu\nu} = \tfrac{1}{2}(\Gamma^\lambda_{\mu\nu} + \Gamma^\lambda_{\nu\mu}) + \tfrac{1}{2}(\Gamma^\lambda_{\mu\nu} - \Gamma^\lambda_{\nu\mu}). \qquad (5.14.2)$$

The second term on the right-hand side of (5.14.2) contributes nothing to the sum in (5.14.1), since the individual terms cancel in pairs when written out explicitly. Consequently a non-symmetric $\Gamma^\lambda_{\mu\nu}$ may be replaced by the symmetric expression $\tfrac{1}{2}(\Gamma^\lambda_{\mu\nu} + \Gamma^\lambda_{\nu\mu})$. So, without loss of generality, we can start with a symmetric $\Gamma^\lambda_{\mu\nu}$.

Now note that the transformation inverse to (5.14.1) is determined, by successive approximation, to be

$$x^\lambda - x^\lambda_P = (\xi^\lambda - \xi^\lambda_P) - \tfrac{1}{2}\Gamma^\lambda_{\mu\nu}(\xi^\mu - \xi^\mu_P)(\xi^\nu - \xi^\nu_P) + \ldots$$

to second order. Therefore

$$dx^\lambda \approx d\xi^\lambda - \tfrac{1}{2}\Gamma^\lambda_{\mu\nu}\,[d\xi^\mu\,(\xi^\nu - \xi^\nu_P) + (\xi^\mu - \xi^\mu_P)\,d\xi^\nu],$$

and

$$\begin{aligned}
ds^2 &= g_{\mu\nu}\left[x^\lambda_P + (x^\lambda - x^\lambda_P)\right] dx^\mu\, dx^\nu \\
&\approx \Big\{ g_{\mu\nu}(P) + \frac{\partial g_{\mu\nu}}{\partial x^\lambda}(P)(\xi^\lambda - \xi^\lambda_P) - \Gamma^\lambda_{\mu\rho}\, g_{\lambda\nu}(P)\,(\xi^\rho - \xi^\rho_P) - \Gamma^\lambda_{\rho\nu}\, g_{\mu\lambda}(P)\,(\xi^\rho - \xi^\rho_P) \Big\} \\
&\quad \times d\xi^\mu\, d\xi^\nu
\end{aligned}$$

on using the Taylor expansion of $g_{\mu\nu}$ about (x^λ_P), substituting for x^λ and dx^λ in terms of the ξ-coordinates, and collecting terms. Note that the summation convention works only if indices are repeated no more than twice, so a certain amount of care is required in choosing the labels to this end. Requiring that the first-order terms be zero gives

$$\frac{\partial g_{\mu\nu}}{\partial x^\rho} - \Gamma^\lambda_{\mu\rho}\,g_{\lambda\nu} - \Gamma^\lambda_{\rho\nu}\,g_{\mu\lambda} = 0$$

at P, which is satisfied if

$$g_{\lambda\nu}\,\Gamma^\lambda_{\mu\rho} = \tfrac{1}{2}\left\{ \frac{\partial g_{\mu\nu}}{\partial x^\rho} + \frac{\partial g_{\nu\rho}}{\partial x^\mu} - \frac{\partial g_{\mu\rho}}{\partial x^\nu} \right\}. \qquad (5.14.3)$$

To get $\Gamma^\lambda_{\mu\rho}$ itself, we have to multiply by the matrix inverse of $(g_{\lambda\nu})$, which is written as $(g^{\mu\lambda})$, for reasons to be discussed later. The components of the identity matrix, $\mathbb{1}$, are conventionally written as δ^λ_μ, called the 'Kroneker delta', so

$$\delta^\lambda_\mu = 1 \qquad \text{if } \lambda = \mu$$
$$= 0 \qquad \text{if } \lambda \neq \mu.$$

Therefore

$$g^{\mu\lambda} g_{\lambda\nu} = \delta^\mu_\nu$$

by the definition of an inverse. Taking the λ component of the identity $\mathbb{1}\mathbf{x} = \mathbf{x}$ gives $\delta^\lambda_\mu x^\mu = x^\lambda$, which is the basic 'index substitution' property of δ^λ_μ required for subsequent manipulations. Multiplying (5.14.3) through by $g^{\sigma\nu}$ gives

$$g^{\sigma\nu} g_{\nu\lambda} \Gamma^\lambda_{\mu\rho} = \delta^\sigma_\lambda \Gamma^\lambda_{\mu\rho} = \Gamma^\sigma_{\mu\rho} = \tfrac{1}{2} g^{\sigma\lambda} \left(\frac{\partial g_{\mu\lambda}}{\partial x^\rho} + \frac{\partial g_{\lambda\rho}}{\partial x^\mu} - \frac{\partial g_{\mu\rho}}{\partial x^\lambda} \right), \qquad (5.14.4)$$

which is the required result. The $\Gamma^\lambda_{\mu\nu}$ are called the components of an *affine connection*. This result provides a general method for obtaining normal coordinates and obviates the need to guess a suitable transformation in each particular case. A further standard linear transformation reduces $g_{\mu\nu}(P)$ to the canonical form, $\eta_{\mu\nu}$.

In ξ-coordinates in a neighbourhood of P, special relativity is valid to first order in ξ. Now in special relativity there is a simple way of labelling points in space–time which exactly parallels that used in ordinary Euclidean geometry. We fix an origin and from it draw vectors (directed line segments) to each event. Certain of these are chosen as base vectors and constitute a *frame of reference*. Call these $\mathbf{e}_{(\lambda)}$ with $\lambda = 0, 1, 2, 3$ in space–time and $\lambda = 1, 2$ in the two-dimensional Euclidean analogue of figure 5.1. Other vectors are specified, and, in practice, measured in terms of their components relative to this reference frame. The events at the tips of the vectors are labelled by these components as coordinates. There are two ways of specifying these components as shown clearly in the diagram (figure 5.1). We can

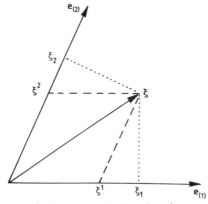

Figure 5.1. The covariant components (ξ_1, ξ_2) and contravariant components (ξ^1, ξ^2) of a vector ξ in the reference frame $(\mathbf{e}_{(1)}, \mathbf{e}_{(2)})$.

use either the perpendicular projection of the vector ξ on to the axes, namely $\xi_\lambda = \xi \cdot e_{(\lambda)}$ or the parallel projection, written ξ^λ, which is defined by $\xi = \xi^\lambda e_{(\lambda)}$. Both methods are equally valid but the numerical values of the coordinates obtained are not the same in the two cases, except in very special circumstances. For, we have

$$\xi_\lambda = \xi \cdot e_{(\lambda)} = \xi^\mu \, e_{(\mu)} \cdot e_{(\lambda)} \neq \xi^\lambda \qquad (5.14.5)$$

unless $e_{(\mu)} \cdot e_{(\lambda)} = \delta^\lambda_\mu$, i.e. unless we are using Cartesian frames of orthogonal unit vectors in Euclidean space.

In special relativity the square of the proper distance from the origin to a point with Minkowski coordinates (ξ^μ) is, by definition, $\eta_{\mu\nu}\xi^\mu\xi^\nu$. But it must be given also by the scalar product of the vector ξ with itself: so

$$\eta_{\mu\nu}\xi^\mu\xi^\nu = \xi \cdot \xi = (e_{(\mu)}\xi^\mu) \cdot (e_{(\nu)}\xi^\nu) = e_{(\mu)} \cdot e_{(\nu)}\xi^\mu\xi^\nu.$$

Therefore

$$\eta_{\mu\nu} = e_{(\mu)} \cdot e_{(\nu)}$$

in a Minkowski frame of reference. In the literature the set of vectors $e_{(\mu)}$, $(\mu = 0, 1, 2, 3)$ is often referred to as a reference *tetrad* or *vierbein*.

Suppose that instead of the reference frame $e_{(\mu)}$ we choose a new set of axes, $e'_{(\mu)}$. Since any vector can be written as a linear sum of basis vectors with suitable coefficients, we must have

$$e'_{(\mu)} = \ell_\mu{}^\nu e_{(\nu)}$$

for some matrix of coefficients $\ell_\mu{}^\nu$ (with rows labelled by μ and columns by ν). For example, $\ell_\mu{}^\nu$ might represent a Lorentz transformation to a moving observer.

The transformation of the coordinate components can now be determined; we have

$$\xi'_\lambda = \xi \cdot e'_{(\lambda)} = \ell_\lambda{}^\nu \xi \cdot e_{(\nu)} = \ell_\lambda{}^\nu \xi_\nu, \qquad (5.14.6)$$

and

$$\xi^\nu e_{(\nu)} = \xi = \xi'^\mu e'_{(\mu)} = \xi'^\mu \ell_\mu{}^\nu e_{(\nu)}. \qquad (5.14.7)$$

Therefore

$$\xi^\nu = \xi'^\mu \ell_\mu{}^\nu$$

which is not the same as (5.14.6). From (5.14.7) we have

$$(\ell^{-1})_\nu{}^\lambda \xi^\nu = (\ell^{-1})_\nu{}^\lambda \ell_\mu{}^\nu \xi'^\mu = (\ell^{-1}\ell)_\mu{}^\lambda \xi'^\mu = \delta^\lambda_\mu \xi'^\mu = \xi'^\lambda$$

Therefore, to get from the unprimed to the primed coordinates in one case we use the transformation matrix ℓ, and in the other case its inverse, ℓ^{-1}. The physical object, the directed line segment ξ, can be represented by these two types of components, called covariant components (ξ_λ), and contravariant components (ξ^λ)

which transform in different ways from one reference frame to another. The two types of components are related by (5.14.5), which, in a Minkowski frame is just

$$\xi_\lambda = \eta_{\mu\lambda}\,\xi^\mu$$

and its inverse, written

$$\xi^\mu = \eta^{\mu\lambda}\,\xi_\lambda.$$

Under a transformation of coordinates the proper distance of an event from the origin is unchanged, but in terms of the new coordinates (ξ'^μ) its square is

$$\xi \cdot \xi = \xi'^\mu \xi'^\nu \mathbf{e}'_{(\mu)} \cdot \mathbf{e}'_{(\nu)} = \xi'^\mu \xi'^\nu \ell_\mu{}^\rho \ell_\nu{}^\sigma \mathbf{e}_{(\rho)} \cdot \mathbf{e}_{(\sigma)}.$$

If we start from a Minkowski frame, $\mathbf{e}_{(\mu)}$, we define

$$g_{\mu\nu} = \ell_\mu{}^\rho \ell_\nu{}^\sigma \eta_{\rho\sigma}$$

and obtain

$$ds^2 = g_{\mu\nu}\xi'^\mu\xi'^\nu.$$

Therefore in a general reference frame at P, equation (5.14.5) becomes

$$\xi_\lambda = g_{\mu\nu}\xi^\mu$$

with its inverse

$$\xi^\mu = g^{\mu\lambda}\xi_\lambda.$$

This is summarised by saying that the metric can be used to 'raise' and 'lower' the indices of vectors.

For consistency, we must show that $(g^{\mu\nu})$, as defined by raising both indices on $(g_{\mu\nu})$, really is the inverse of $(g_{\mu\nu})$. In matrix notation we have, obviously

$$g^{-1}gg = g.$$

By using two matrices, $(g_{\mu\nu})$, to lower the two indices on $(g^{\mu\nu})$, we have also

$$g^{\mu\nu}g_{\mu\lambda}g_{\nu\rho} = g_{\lambda\rho}.$$

Comparing these equations we see that the components of g^{-1} are $(g^{\mu\nu})$.

The components of physical objects may occur naturally in either covariant or contravariant form. Suppose $\phi(\xi)$ is the value of a physical field near P and form the gradient field $\partial\phi/\partial\xi^\lambda$. To find out what sort of components these are we investigate the transformation law to a new coordinate system. So let $\xi^\lambda = \xi'^\nu \ell_\nu{}^\lambda$ define new coordinates (ξ'^λ). Then

$$d\phi = (\partial\phi/\partial\xi^\lambda)\,d\xi^\lambda = (\partial\phi/\partial\xi^\lambda)\,d\xi'^\nu \ell_\nu{}^\lambda;$$

but also

$$d\phi = (\partial\phi/\partial\xi'^\nu)\,d\xi'^\nu;$$

112

therefore,

$$\partial\phi/\partial\xi'^{\nu} = \ell_{\nu}{}^{\lambda}\,\partial\phi/\partial\xi^{\lambda}.$$

This is like (5.14.6), so these are covariant components of the gradient. Note that this argument also demonstrates that the gradient is a vector; this follows from the transformation law, which shows that the gradient transforms like a directed line segment, so can be represented by a position vector. Anything which does not transform like this is not a vector because it cannot be equated to a directed line segment in an arbitrary coordinate system. Thus the physical character of an object is contained in the transformation law of its components.

Objects which transform in different ways from vectors are important and are therefore given names. For example, an object transforming according to

$$T'^{\mu\nu}{}_{\alpha} = (\ell^{-1})_{\lambda}{}^{\mu}\,(\ell^{-1})_{\rho}{}^{\nu}\,\ell_{\alpha}{}^{\beta}\,T^{\lambda\rho}{}_{\beta}$$

is a mixed tensor of covariant rank 1 and contravariant rank 2. Thus the partial derivatives $\partial^2\phi/\partial\xi^{\lambda}\,\partial\xi^{\mu}$ are the covariant components of a second-rank tensor. The key point is that both sides of an equation of physical significance must transform in the same way. So from the vector ξ^{λ} we can form the quantities

$$\xi_{\lambda}\xi_{\lambda}, \qquad \xi_{\lambda}\xi^{\lambda} \equiv \eta_{\lambda\mu}\xi^{\lambda}\xi^{\mu} \equiv \eta^{\lambda\mu}\xi_{\lambda}\xi_{\mu}, \qquad \xi_{\lambda}\xi_{\mu}$$

(with summation over repeated indices). The first is unlikely to be of much significance, since it does not transform in a simple way. The second is a tensor of rank 0 (otherwise called a 'scalar'), and has the same value in all coordinate systems; we have met an example in the square of the proper distance. The last are the covariant components of a second-rank tensor, which could be of some use.

So far we have concentrated on special relativity, but these same considerations apply to first order in normal coordinates in a small region of the curved space–times of the general theory. To develop our promised, more powerful theory we need to extend these ideas to general coordinate systems. The transformations we shall want to consider can vary from point to point, so the $\ell_{\mu}{}^{\nu}$ may be functions of position. By differentiating $\xi^{\lambda} = \xi'^{\mu}\ell_{\mu}{}^{\lambda}$ we get $\ell_{\mu}{}^{\lambda} = \partial\xi^{\lambda}/\partial\xi'^{\mu} = \partial x^{\lambda}/\partial x'^{\mu}$ at P, the origin of normal coordinates. Therefore the transformation law for the contravariant components of a vector can be written as

$$A'^{\mu} = \frac{\partial x'^{\mu}}{\partial x^{\nu}}\,A^{\nu},$$

which therefore defines a vector at P in arbitrary coordinate systems in a general space–time. But since P is arbitrary, this form is valid everywhere, and we need make no reference to special coordinate systems at particular points! This is the first illustration of the method of extending the theory to be pursued in the rest of this section. Note that the form of this equation explains the choice of superscript labels for the x's (which are not contravariant vectors but arbitrary coordinates).

113

There is an obvious extension to covariant components

$$A'_\mu = \frac{\partial x^\nu}{\partial x'^\mu} A_\nu$$

and to general tensors. As a matter of terminology, tensors defined as functions of position and time are called tensor 'fields' (or, in special cases, scalar fields and vector fields).

Consider next the transcription of the equation of geodesics from normal coordinates at a point P to a generally valid form. In normal coordinates at P geodesics are straight lines given by

$$\frac{d^2 \xi^\mu}{d\tau^2} = 0.$$

To transform this to general coordinates we need to keep those second-order terms in the transformation (5.14.1) that do not vanish in the limit $\xi^\lambda \to \xi^\lambda_P$, thus we get

$$0 = \frac{d^2 \xi^\mu}{d\tau^2} = \frac{d^2 x^\mu}{d\tau^2} + \Gamma^\mu_{\nu\rho} \frac{dx^\nu}{d\tau} \frac{dx^\rho}{d\tau} \tag{5.14.8}$$

at P. But since P is arbitrary this must be the relevant equation everywhere. This provides a direct derivation of the general form of the geodesic equation by essentially the same argument as was used previously in particular cases. For this comparison, note that in the example of §5.11 the time coordinate ξ^0 was called λ.

An alternative description is provided by means of the concept of a *covariant derivative*, which is of considerable importance in more advanced treatments. In a general coordinate system the rate of change of the components of a vector with position and time depends as much on the arbitrary choice of coordinates as on any physical changes in the field: stretch the space coordinates and the rate of change is slow, squash them to speed it up. A physically meaningful rate of change is obtained by using a reference frame set up in normal coordinates (ξ^λ) like those in our discussion of special relativity above. We then transcribe this to general coordinates (x^λ) to obtain the covariant derivative. Thus, we have

$$\frac{\partial A^\mu(\xi)}{\partial \xi^\rho} = \frac{\partial x^\lambda}{\partial \xi^\rho} \frac{\partial}{\partial x^\lambda} \left(\frac{\partial \xi^\mu}{\partial x^\nu} A^\nu(x) \right),$$

using the transformation law for vectors and the chain rule for partial derivatives. Using the transformation formula (5.14.1) gives, *at P*

$$\left\{ \frac{\partial A^\mu(\xi)}{\partial \xi^\rho} \right\}_P = \left\{ \frac{\partial A^\mu}{\partial x^\rho}(x) + \frac{\partial^2 \xi^\mu}{\partial x^\rho \partial x^\nu} A^\nu(x) \right\}_P = \left\{ \frac{\partial A^\mu}{\partial x^\rho} + \Gamma^\mu_{\rho\nu} A^\nu \right\}_P.$$

The final expression gives a physically meaningful rate of change of A^μ in arbitrary coordinates at any point. It is called the covariant derivative of A^μ, and is written DA^μ/Dx^ν. It can be shown that this covariant derivative transforms as a tensor (of covariant rank one, contravariant rank one), whereas neither of its two constituent terms is a tensor. As an application note that the geodesic equation can be written

$$0 = \frac{D}{D\tau}\left(\frac{dx^\mu}{d\tau}\right) \equiv \frac{dx^\lambda}{d\tau}\frac{D}{Dx^\lambda}\left(\frac{dx^\mu}{d\tau}\right).$$

We can now show that a geodesic is a curve which extremises the proper time. For the Euler–Lagrange equations of the variational principle

$$\delta \int d\tau = \delta \int \left\{ g_{\mu\nu}\frac{dx^\mu}{d\tau}\frac{dx^\nu}{d\tau}\right\}^{1/2} d\tau = 0$$

can be shown to be equivalent to the geodesic equation (5.14.8) on using the explicit form for $\Gamma^\lambda_{\mu\nu}$ from (5.14.4). Provided that the parameter τ is taken to be proper time, an equivalent variational principle, which is simpler to use for explicit calculations, is

$$\delta \int g_{\mu\nu}\frac{dx^\mu}{d\tau}\frac{dx^\nu}{d\tau} d\tau = 0.$$

This can be verified again by explicit computation of the Euler–Lagrange equations.

Finally, we turn to the second-order terms in the expansion of the metric in normal coordinates. At this point it is necessary to admit to a semantic inaccuracy in the preceding discussion. So far we have referred to 'normal coordinates' as if we were using some uniquely prescribed system. In fact, this is not the case, since the inclusion of higher-order terms in the transformation (5.14.1) provides an infinite variety of choices for 'normal coordinates'. All of these would be equally suitable up to now. However, the precise values of the second-order deviation of the metric coefficients from constants in a neighbourhood of P depend on the choice of the third-order terms in the transformation. To make the discussion precise we have to specify this. Thus, we shall choose new 'normal coordinates' to be defined by the transformation

$$\xi^\lambda - \xi^\lambda_P = x^\lambda - x^\lambda_P + \tfrac{1}{2}\Gamma^\lambda_{\mu\nu}(x^\mu - x^\mu_P)(x^\nu - x^\nu_P) + \tfrac{1}{6}\Gamma^\lambda_{\mu\nu}\Gamma^\mu_{\rho\sigma}(x^\nu - x^\nu_P)(x^\rho - x^\rho_P)(x^\sigma - x^\sigma_P)$$

with $\Gamma^\lambda_{\mu\nu}$ given, as usual, by (5.14.4) *evaluated at P*. This choice has the advantage that the inverse takes the particularly simple form

$$x^\lambda - x^\lambda_P = \xi^\lambda - \xi^\lambda_P - \tfrac{1}{2}\Gamma^\lambda_{\mu\nu}(\xi^\mu - \xi^\mu_P)(\xi^\nu - \xi^\nu_P) + \text{fourth-order terms,}$$

with no third-order terms at all. Note that normal coordinates are usually fixed on more obviously physical grounds, and that those in use in the literature are not necessarily the same as the ones chosen here. However, the result we obtain will be the standard one.

We seek the form of the metric in ξ-coordinates in a neighbourhood of P, retaining now terms of second order. Thus, we have

$$dx^\lambda = d\xi^\lambda - \Gamma^\lambda_{\mu\nu}d\xi^\mu(\xi^\nu - \xi^\nu_P)$$

115

and, therefore,

$$ds^2 = g_{\lambda\mu} \, dx^\lambda \, dx^\mu \approx \left\{ g_{\lambda\mu}(P) + \frac{\partial g_{\lambda\mu}}{\partial x^\gamma} [(\xi^\gamma - \xi_P^\gamma) - \Gamma_{\delta\epsilon}^\gamma (\xi^\delta - \xi_P^\delta)(\xi^\epsilon - \xi_P^\epsilon)] \right.$$

$$+ \frac{1}{2} \frac{\partial^2 g_{\lambda\mu}}{\partial x^\gamma \partial x^\delta} (\xi^\gamma - \xi_P^\gamma)(\xi^\delta - \xi_P^\delta) \Big\}$$

$$\times \, [d\xi^\lambda - \Gamma_{\alpha\beta}^\lambda \, d\xi^\alpha (\xi^\beta - \xi_P^\beta)] \, [d\xi^\mu - \Gamma_{\rho\sigma}^\mu \, d\xi^\rho (\xi^\sigma - \xi_P^\sigma)]$$

$$\approx \left\{ g_{\lambda\mu}(P) + \left[\frac{1}{2} \frac{\partial^2 g_{\lambda\mu}}{\partial x^\nu \partial x^\rho} - \Gamma_{\nu\rho}^\alpha \frac{\partial g_{\mu\lambda}}{\partial x^\alpha} - \Gamma_{\mu\nu}^\alpha \frac{\partial g_{\alpha\lambda}}{\partial x^\rho} - \Gamma_{\lambda\rho}^\alpha \frac{\partial g_{\mu\alpha}}{\partial x^\nu} \right. \right.$$

$$\left. \left. + \Gamma_{\mu\nu}^\alpha \Gamma_{\lambda\rho}^\beta \, g_{\alpha\beta} \right]_P (\xi^\rho - \xi_P^\rho)(\xi^\nu - \xi_P^\nu) \right\} \, d\xi^\lambda \, d\xi^\mu. \tag{5.14.9}$$

A general form for this expression would be

$$ds^2 \approx [g_{\lambda\mu}(P) + g_{\lambda\mu\nu\rho}(P) (\xi^\nu - \xi_P^\nu)(\xi^\rho - \xi_P^\rho)] \, d\xi^\lambda \, d\xi^\mu. \tag{5.14.10}$$

The quantity $g_{\lambda\mu\nu\rho}$ may be taken to be symmetric in each pair of indices, (λ,μ) and (ν,ρ), by an extension of the argument used above to justify the symmetry of $\Gamma_{\mu\nu}^\lambda$. Thus, we may impose

$$g_{\lambda\mu\nu\rho} = g_{\mu\lambda\nu\rho}; \qquad g_{\lambda\mu\nu\rho} = g_{\lambda\mu\rho\nu}$$

in comparing (5.14.9) and (5.14.10) to get $g_{\lambda\mu\nu\rho}$.

Rather than $(g_{\lambda\mu\nu\rho})$ itself, we choose certain linear combinations of the components, partly because we want the relevant quantity to turn out to be a tensor, and partly on historical grounds. Therefore, we define the Riemann curvature tensor

$$R_{\lambda\rho\nu\mu} = -g_{\lambda\mu\nu\rho} + g_{\rho\mu\nu\lambda} + g_{\lambda\nu\mu\rho} - g_{\rho\nu\mu\lambda}, \tag{5.14.11}$$

and compute this to be

$$R_{\lambda\rho\nu\mu} = -\frac{1}{2} \left(\frac{\partial^2 g_{\lambda\mu}}{\partial x^\nu \partial x^\rho} - \frac{\partial^2 g_{\rho\mu}}{\partial x^\nu \partial x^\lambda} - \frac{\partial^2 g_{\lambda\nu}}{\partial x^\mu \partial x^\rho} + \frac{\partial^2 g_{\rho\nu}}{\partial x^\mu \partial x^\lambda} \right) + g_{\alpha\beta} (\Gamma_{\mu\rho}^\alpha \, \Gamma_{\nu\lambda}^\beta - \Gamma_{\mu\lambda}^\alpha \, \Gamma_{\nu\rho}^\beta).$$
$$\tag{5.14.12}$$

With this approach it is straightforward, although tedious, to show that $R_{\lambda\mu\nu\rho}$ transforms like a (fourth-rank covariant) tensor. If $R_{\lambda\mu\nu\rho}$ is itself expressed in terms of normal coordinates at P, by the appropriate transformation, the terms involving Γ vanish, and $R_{\lambda\mu\nu\rho}$ is seen to be just a combination of second derivatives of the metric tensor, arranged to have certain symmetries.

The Riemannian curvature represents the departure from 'flatness' due to tidal gravitational forces, and controls the deviation of geodesics. It is essentially the Riemann tensor that is related to the presence of energy and momentum in space-time through Einstein's equations (§6.6).

6. *Cosmological Models*

6.1. Mach's Principle

In cosmology, as in most branches of science, the historical record is an unfaithful guide to the logic of scientific discovery. From the logical point of view the development of general relativity emerges from a reformulation of the Newtonian theory of gravity in response to criticisms of the conceptual basis of that theory, criticisms which were formulated in detail only after Einstein had succeeded in obtaining his general theory (Raine and Heller 1981). But in historical reality, Einstein proceeded on the basis of an altogether different critical analysis of Newtonian dynamics, and was led, via the Equivalence Principle, directly from the special theory of relativity to the general theory. Thus the historical revolution is underwritten by a logical evolution. The critique of Newtonian theory which influenced Einstein was due to the Austrian physicist Ernst Mach (1883). To the main ideas which Einstein loosely drew from Mach, he gave the name Mach's Principle. In a logical presentation of the general theory it is entirely possible to avoid mention of this Principle, as indeed we have done. It is nevertheless relevant for us here, since it led not only to general relativity but also to cosmology, and may have a bearing on the isotropy of the Universe (Chapter 10).

A precise statement of what Mach's Principle is, or should be, is to a certain extent a matter of opinion, since Einstein never provided one. However, the idea that the Principle is to encompass is quite straightforward. From a kinematical point of view there can be motion only of bodies relative to other bodies. For example, there should be no difference between a rotating Earth in a stationary universe, and an Earth at rest with the stars in rotation about it. The problem is that from the point of view of Newtonian dynamics there is a considerable differ-ence. If the Earth were dynamically non-rotating, with the stars and galaxies moving round it, there would be no flattening of the poles, no rotation of the plane of the Foucault pendulum; in short, an Earth-bound observer would feel no inertial forces of the centrifugal or Coriolis type. Of course Newton solved the problem by hypothesising an absolute space with respect to which true dynamical motion could be measured, and with respect to which the Earth could be deemed to be truly rotating. In practical applications, we solve the problem by appealing to observations to tell us that by taking dynamical motion to be measured relative to the 'fixed' stars we shall obtain the correct results from the theory. However, Newtonian theory gives no reason for doing this, other than the fact that it works. Mach's Principle asserts that the problem should be solved by the construction of a theory of dynamics in which there really would be no difference between a universe in which the Earth rotates relative to the stars and one in which the stars rotate about the Earth: dynamics should involve only this relative motion.

This is what Einstein set out to achieve. It is clear to us now, although it was not, apparently, to Mach, that we need at the very least a dynamical theory in which the rotating stars can exert forces on the Earth by virtue of their rotational acceleration. These forces must then be exactly the same as the centrifugal and Coriolis forces obtained by starting with the stars at rest and transferring to the rotationally accelerated frame of the Earth. Now, according to the Equivalence Principle, we cannot distinguish between forces obtained by acceleration and forces due to gravitation. Thus, Einstein was led to suggest gravity as the mechanism by which the rotating stars flatten the poles of the Earth.

Clearly Newtonian gravity will not do for this; the Newtonian gravitational force generated by a body depends only on its mass and its distance, and is independent of its state of motion. We need a theory with extra components to the gravitational force, but we have seen that this is just what a relativistic theory of gravity has to provide. The only question, then, is whether in general relativity these extra forces are always of exactly the right magnitude to ensure that kinematically equivalent motions are also dynamically equivalent.

In general relativity the motion of a body subject to the influence of gravity only is determined by the metric coefficients. Therefore, we can rephrase this question of the relation of the gravitational forces to the distribution and motion of gravitating matter; we consider instead the relation of the geometry of space–time to the matter it contains. This relation is, as we have seen, provided by Einstein's field equations. The field equations themselves do not tell us what the geometry of space–time must be, because they do not tell us what the distribution of matter must be. They do tell us which possible geometries can be associated with which distributions of matter.

Mach's Principle involves the knowledge of the whole Universe, since all the matter is supposed to contribute to the geometry in any region. To investigate Mach's Principle therefore leads immediately to cosmology. We have to make some approximation to the distribution of matter on a large scale. We then solve the field equations to find the corresponding geometry, and from this the motion of bodies. In this way we construct a *model universe* – a possible universe according to the laws of physics. We hope that the model bears as close a relation to the dynamics of the real Universe as does the approximation of the large-scale matter distribution to the true distribution of galaxies. If general relativity were success-fully to incorporate Mach's Principle, we should find that in *all* of these possible model universes the forces of the rotating stars should be sufficient to cause the flattening of the poles of the Earth. In the theory of relativity, just as in Newtonian theory, this flattening can be predicted correctly, since the calculations can be performed on the assumption that the Earth is rotating relative to a dynamical inertial frame. Unfortunately, when the calculations are made *ab initio* for an Earth at rest, only in exceptional models are the forces due to the then rotating stars sufficient to produce the observed results.

To see this, consider a model consisting of an Earth and only a handful of distant stars. It turns out that this is permitted by the theory, although, of course, it does not look anything like our real Universe. The rotation of these stars sufficiently distant from the Earth has an almost negligible effect. Even according to general relativity, the flattening of the Earth's poles in such a universe would occur, and would be produced by rotation relative to the ghost of absolute space, not relative to the essentially irrelevant stars. Conversely, according to the theory, universes filled with matter everywhere are possible in which the daily swirling of all the stars produces no flattening effect at all on the Earth! Also, at the other extreme, a model universe is possible which is devoid of all matter and yet in which the motion of test bodies would be in exact agreement with our laboratory observations. As far as laboratory dynamics is concerned, it would appear that one is as free to remove the stars in relativity as one is in Newtonian theory.

At first Einstein hoped to solve this problem by modifying the field equations he had postulated. There was an additional, and, ironically, what was for him probably a more compelling reason for this modification: Einstein was having difficulty in deriving any cosmological models at all from his original field equations! He introduced into the equations an extra term, which he called the cosmological term, thereby creating a new constant of nature, the cosmological constant, Λ. This term acts as a cosmic pressure exerted by space itself, thereby modifying the effect of gravity over large distances. This is superficially similar to the modification of the Newtonian inverse square law in equation (5.3.2), although the effect of the two Λ-terms is quite different. The introduction of Λ produces a finite range for gravity in the Newtonian case, and a cosmic repulsion in general relativity. With the new constant, Einstein obtained the cosmological model to be discussed in §6.2.

Why did Einstein experience difficulty in constructing a cosmological model from the original equations? It turns out these equations are incompatible with a static universe (§6.7). In this respect the consequences of relativity are no different from those of Newtonian theory (§5.3), but what then becomes of Mach's Principle in the modified theory? Unfortunately, the conclusions are unchanged: de Sitter was able to construct a cosmological model, in accord with the modified theory, which again contains no matter to produce gravitational effects, but in which the motion of test bodies agrees with local observations. The stars are equally dispensable with or without the Λ-term. Einstein was later to describe the cosmological constant, somewhat unjustly, as the biggest blunder of his life.

What then is the role of Mach's Principle in the theory of relativity, apart from that of an historical fulcrum? There are three possible responses to this question.

The simplest is to bury it with all the other ill-articulated pseudo-philosophical gropings that often accompany the emergence of new physical insights. The most drastic is to abandon Einstein's theory, either entirely or in part, as having failed to embody the true guiding spirit of Mach, and to try to do better. So far this approach has produced only much that is distinctly worse. It appears immediately less promising in the light of the increasingly good agreement that is being achieved between the predictions of general relativity and a wide range of observations. An intermediate response is to turn the problem round. If general relativity allows both models that do and models that do not satisfy Mach's Principle, then we can insist on the validity of the Principle to *select* only the former as physically reasonable models for the Universe. In this way one might hope to explain to some extent why the Universe is as it is. We shall take this up again in Chapter 10.

There is, unfortunately, also a fourth possibility. According to this one may adopt any combination of the three aforementioned attitudes, together with any amount of reinterpretation of what it is that Mach's Principle ought to say. This has generated much confusion in the literature, and a flourishing sub-culture.

6.2. The Einstein Model

In our present state of knowledge a non-zero but small cosmological constant is compatible with both theory and observation. However, for the most part, a small non-zero Λ makes little qualitative difference to the theory, but does add technical complications. For most of our discussion, we shall set $\Lambda = 0$, adopting the attitude that a new constant of nature should be introduced only when it is forced on us by the observations. Nevertheless, for the purpose of illustration in this section, we shall consider Einstein's first attempt to construct a model of the Universe, for which he took $\Lambda \neq 0$. In this case there is no analogous model with $\Lambda = 0$.

For the construction of the first model of the large-scale structure of the Universe in 1917, Einstein made two assumptions. The first was that matter is distributed throughout space with a density that is independent of position. There are two senses in which this appears at first to be unrealistic. First, in the real Universe matter is at present agglomerated into galaxies, and throughout most of space the density of matter is virtually zero. This is a purely technical complication. As discussed in Chapter 1, we simply imagine the galaxies to be like the atoms of a gas; on a large enough scale we can consider an average density of matter and treat both the gas and the Universe as continuous fluids. If the averaging is performed over the scale of many clusters we can consider the fluid to have uniform density. Although we know that there are no preferred clustering scales (Chapter 1), it is often useful, for the purpose of visualisation, to think of clusters of galaxies as the analogues of atoms in a gas. Secondly, the real Universe, as Einstein knew it in 1917, consisted of only the finite system of stars that make up our Galaxy. Until perhaps 1924, when Hubble established a distance to the Andromeda Nebula, the extension of the Universe beyond our Galaxy was pure speculation. The assumption of a uniform Universe with no edge was therefore initially to be justified by mathematical convenience. Today, of course, this second objection has less weight. We have seen that for the visible Universe the assumption of uniformity is justified to an extent well beyond the hopes of even the most optimistic of the early cosmologists. On the other hand, the assumption of uniformity for the parts we cannot see goes well beyond the laws of evidence. It is a consequence of the 'Cosmological Principle', which we discuss in §6.3.

Einstein's second assumption was that the Universe is constant in time; again this is to be taken to apply in an average sense. One might object that since the Universe clearly contains radiation and matter not in equilibrium, it must be evolving to equilibrium, and cannot be static. The interconversion of forms of energy from matter to radiation certainly does have a detailed effect on the equation of state of the gravitating mass–energy of the Universe. But this is not the most important aspect of the objection. The Universe cannot be both infinitely old and not in equilibrium, and this form of the argument points immediately to a non-static Universe. Ironically, in this case, Einstein accepted the direct observations which showed no evidence for anything other than a static world. To be fair here, it is not unreasonable to have hoped that a static model would turn out to be a good approximation to one that is changing only slowly. It turns out that this is not really the case.

One now makes the further assumption that the particles of the matter fluid (clusters of galaxies) have no random motions, and so exert zero pressure. Then an observer falling freely with the matter in the Einstein universe can be uniquely distinguished at each point. The time measured by such observers can be used as the coordinate t. In this way we imagine the universe to be filled by a set of 'fundamental' observers. Mathematically one can imagine a picture of clocks attached to each point of the fluid; physically one can imagine each clock attached to a cluster of galaxies. The clocks can be synchronised in such a way that the geometry of Einstein's model is described by the metric

$$-d\tau^2 = -dt^2 + a_E^2 [dr^2 + \sin^2 r(d\theta^2 + \sin^2\theta\, d\phi^2)], \qquad (6.2.1)$$

where a_E is a constant. It turns out that the Einstein field equations yield a value for a_E in terms of the density of matter, ρ, in the Universe, namely

$$a_E = \left(\frac{4\pi G}{c^2}\rho\right)^{-1/2},$$

provided the cosmological constant Λ is chosen as $\Lambda = a_E^{-2}$ (§6.7).

The formula for the metric itself tells us what the coordinates mean, at least throughout some finite region of space–time in which the expression is defined. The expression $d\theta^2 + \sin^2\theta\, d\phi^2$ looks like the distance between two points on the surface of a sphere labelled by polar coordinates, and must therefore be exactly that. The factor multiplying this is $a_E^2 \sin^2 r$, so if we allow θ and ϕ to vary, with r and t held fixed, we generate a sphere of radius $a_E \sin r$. In the surface t = constant,

ϕ = constant, the distance between points with labels differing by dr and $d\theta$ is $a_E^2(dr^2 + \sin^2 r \, d\theta^2)$. Apart from the change in notation this is again the metric of the surface of a sphere. Since the notation cannot have any physical significance this must again *be* the geometry of the model. Although r looks like it wants to be a radial coordinate running from 0 to ∞ and having units of length, the metric expression tells us it must be an angle running from 0 to π (figure 6.1).

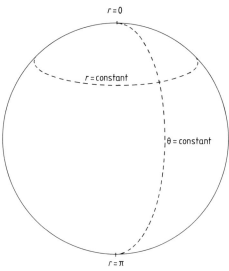

Figure 6.1. The r-θ surface (t = constant, ϕ = constant) in the Einstein model is a sphere of radius a_E. Lines of latitude and longitude on the sphere are labelled by r and θ, with $r = 0$ and $r = \pi$ at the poles and $0 \leqslant \theta < 2\pi$.

The geometry of space at a constant time, t, is that of the surface of a three-dimensional sphere of radius a_E embedded in a four-dimensional Euclidean space. It is not necessary to agonise over a mental picture of this, since the metric expression contains all the relevant information, but it is sometimes useful to bear this description in mind. In particular, the spaces t = constant, all of which have the same geometry since the Einstein universe is static, are finite but unbounded.

In order to draw a space–time diagram of the model we have to suppress a certain amount of information. We restrict ourselves to a representation of the (t, r) plane by plotting events with fixed values of θ and ϕ. Since $0 \leqslant r \leqslant \pi$ at a given time defines a circle, we can represent the history of the points of this circle on the surface of a cylinder. Shown in figure 6.2 are the free-falls, which are the space–time paths of clusters of galaxies, or fundamental observers. Light rays are given by $-dt^2 + a_E^2 \, dr^2 = 0$, and hence by $a_E r = \pm t$. The cylindrical surface can be unwrapped into a plane, and the light rays are obtained as straight lines in this plane.

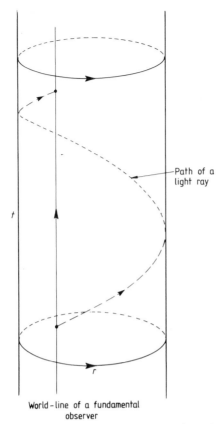

Figure 6.2. Geometry of the space–time of the t, r surface of the Einstein model (θ = constant, ϕ = constant). r runs from 0 to π round the cylinder, and t runs from $-\infty$ to ∞ up it.

The spatial finiteness of the universe now becomes clear. A light ray runs right the way round and back again to its starting point after a finite time. In the absence of intervening absorption, one should be able to see the back side of a galaxy by looking at the antipodal point in the sky. Note how different this global picture is from a space–time diagram of special relativity, even though, in accordance with the Principle of Equivalence, the two are indistinguishable locally.

In fact, the Einstein model cannot be a realistic model of any universe because it is unstable. It describes a system in which the cosmic repulsive force, described by the Λ-term, exactly balances the attractive gravitational force. Any small deviation leads to runaway contraction or expansion. Since fluctuations occur in any physical system, there will be small departures from exact balance and transition to an expanding or collapsing model. It follows that an expanding (or contracting) universe is a prediction of general relativity, which could have been made in advance of the observations.

6.3. Cosmological Principles

If we knew everything about the world we should have no need of principles; but we should have little need of physics either. At the other extreme, assumptions are raised to the status of principles as a defence for our palpable ignorance. We have seen that in order to obtain a cosmological model Einstein had to make an assumption concerning the large-scale distribution of matter. An assumption of this type is often referred to as a *cosmological principle*. In the Einstein model the cosmological principle is that the matter distribution is homogeneous in space, and is static. Within the context of a physical theory, the theory of relativity in this case, this assumption can be shown to lead to consequences not in accordance with observation (the absence of redshifts of distant galaxies, for example), and hence can be shown to be false. In this case the theory as a whole is falsified, but not necessarily the specific cosmological principle. However, we have already seen that, in terms of the observations available at the time, Einstein's principle was literally false and satisfied the demands only of mathematical convenience. If, therefore, we can appeal to observations to yield the distribution of matter in the world, the question immediately arises as to why we should need a cosmological principle at all. That is, even if we do not know everything, might we not for our present purpose know enough?

To understand the negative answer to this question, consider what has come to be known as *the* Cosmological Principle, which is the one we shall adopt throughout this book. This states that the Universe is, on average, homogeneous and isotropic at any one time, by which we mean that it could be deduced to be so by at least one class of observers. We have dropped the requirement that the Universe be static. Note that observers do not necessarily *see* such a universe as homogeneous since the finite velocity of light means that they see different regions at different periods in the past. The theory must provide an unambiguous way of correcting for this effect. The Cosmological Principle will lead us to models of the Universe in reasonable agreement with observation. Unfortunately, within the context of the theory of relativity, it leads also to the conclusion that there is no way of testing the Cosmological Principle separately from the theory as a whole. That is, if the Universe in which we live is indeed like this model, then we cannot directly know it! For we shall find that in such a model at any finite time only a finite region of space is available to direct observation by a given observer, namely a region of space within which the redshifts of galaxies are finite (§8.9). The region accessible to direct observation, at least in principle, becomes larger as time goes on, but it is never the whole Universe. Of course, if the Cosmological Principle is false we may someday hope to discover that by observing the coming into view of the edge of our uniform domain. We shall then have no way of predicting subsequent observations.

Nevertheless, there is a strong expectation that the local observations of isotropy should contribute some degree of testability to the Cosmological Principle. This can be expressed through what has become known as the *Copernican Principle*, which states that we do not occupy a privileged position in the Universe. Unfortunately, the precise meaning of this statement depends on what one considers to be privileged. Nevertheless, it is easy to see that the Copernican Principle, and the local observations of isotropy, imply that the Cosmological Principle holds. For, if matter appears to be distributed isotropically about us, it must be so distributed about all equivalent observers. It follows that the Universe is isotropic about each point, and hence is homogeneous in space.

From this point of view, the Cosmological Principle combines an unverifiable assumption and an observational result. Sometimes it is argued that the Copernican Principle can be given weight on the grounds of probability. This is a difficult argument; one would like to say that it is highly improbable that we should occupy a privileged position near the centre of a lump of an inhomogeneous distribution. Unfortunately, we only call such a position privileged because we assume that it is highly improbable.

It is just the apparently unverifiable element of cosmological principles that led to the tenacity of belief of the proponents of the steady state theory. Here the cosmological principle at stake is the *Perfect Cosmological Principle*, according to which the Universe is not only the same at all places, but at all times. If this principle were not testable, it would be possible to fit the physics to the observations to give a satisfactory theory. To this end Hoyle's 'C-field' was created to create matter to keep the density of the Universe constant. Unfortunately, the detailed physics is irrelevant here: from the Perfect Cosmological Principle, the Principle of Equivalence, and special relativity, there follow predictions which can be shown to be in disagreement with observations (Chapter 7). To this extent this particular cosmological principle *can* be tested and found wanting.

There is another sense in which the Copernican Principle is central to our discussion of cosmology. It can be thought of as the basis for the assumption that the laws of physics here and now are valid everywhere and always, but this assumption, however justified, is not necessarily correct; philosophical principles are hostages to facts. It is therefore certainly worth probing the assumption as far as possible. Indeed a certain amount of attention has been devoted to testing such things as the constancy in space and time of the gravitational constant, G, and of the fine-structure constant $\alpha = e^2/hc$. On the other hand, it is a complete waste of time to contemplate the erection of speculative edifices without this use of the Copernican Principle.

We conclude from this foray into philosophy that some form of cosmological principle is required to do cosmology, and the extent to which it can be tested depends upon what it states. We shall take the view that the Copernican Principle is valid in as strong a form as is necessary to do unambiguous calculations on the basis of our present knowledge.

There is one further comment necessary here. It is possible that the Cosmological Principle holds for us now, but that the Universe came to be this uniform by a process of smearing out of inhomogeneities and anisotropies at some earlier stage. This possibility will be considered in §10.3. It is also possible that the Universe will violate the Cosmological Principle by developing anisotropies in the future. We shall take this up briefly in §§10.1 and 10.4. For the rest we shall assume isotropy at all times, right back to the big bang.

Another feature of the Einstein model, and a consequence, in fact, of the Cosmological Principle, is the existence of a preferred time coordinate. This is the proper time as measured by the 'fundamental observers' introduced in §6.2. We call it *cosmic time*. Any observer moving relative to a fundamental observer will measure a different time.

It is important to be clear that this does not conflict with the requirements of special relativity. Fundamental observers in the Einstein model, or in any model for which the Cosmological Principle holds, can be selected by the criterion that they see the Universe of stars in a particular way, namely as isotropic. Their existence depends therefore on the stars, just as the privileged observers of §3.4 owed their existence to that of the microwave background. Cosmic time is in no way privileged with regard to local laboratory experiments.

In the Einstein universe fundamental observers can be regarded as attached to clusters or to galaxies. This is not an arbitrary choice. The constraining influence is the Einstein field equations. If the galaxies were to be moving relative to fundamental observers this would alter the gravitational fields they generate. Hence the metric would be different, and one would no longer have the Einstein model! Nevertheless, one can envisage the possibility of a consistent model in which there is a relative motion of galaxy clusters and the fundamental observers. If this were non-zero on average, observers attached to clusters would not see other clusters distributed isotropically, nor would their clocks measure cosmic time. Such a model could still be homogeneous in space at constant cosmic time. These spatially homogeneous anisotropic models will be of importance in later chapters. One could, of course, consider models which were not even homogeneous, but this introduces great technical difficulties; the Cosmological Principle is not satisfied, there are no physically privileged fundamental observers, and in general there is no physically distinguished cosmic time.

6.4. Isotropic Geometry

Our first approximation to a model of the Universe will be a space which is homogeneous and isotropic and which changes in time in accordance with the Einstein field equations. One such appropriate model was discussed in Chapter 5: the metric (5.11.1) describes a geometry in which the three-dimensional space at any one cosmic time is Euclidean. Nevertheless, the four-dimensional geometry is 'curved', as is evidenced by the fact that freely falling observers who signal to each other will find increasingly long times between transmission and reception as a result of the divergence of their geodesic world-lines. In this sense the space is expanding. The rate of expansion is given by the function $a(t)$, and this is determined for a particular matter distribution by solving Einstein's equations.

Before we come to these equations, however, the question arises as to whether the metric (5.11.1) is the most general geometry that satisfies the Cosmological Principle. This is equivalent to asking whether space can be anything other than Euclidean and still satisfy the requirements of homogeneity and isotropy. We know that in two dimensions the Euclidean plane is a homogeneous space. So too is the sphere, since it has no preferred points or directions. A 'sphere' of imaginary radius, which cannot be visualised as a surface embedded in three-dimensional Euclidean space, is nevertheless another example. In fact, these three possibilities complete the list of two-dimensional isotropic geometries. For three-dimensional spaces, an analogous result was found independently by Robertson and Walker: there are essentially just three possible isotropic three-dimensional geometries. The metric describing these geometries can be given in the form

$$d\sigma^2 = (1+\tfrac{1}{4}Kr^2)^{-2}\{dx^2+dy^2+dz^2\}. \tag{6.4.1}$$

where K is a constant and $r^2 = x^2+y^2+z^2$. If $K \neq 0$, we can put $\bar{x} = |K|^{1/2}x$ to get

$$d\sigma^2 = |K|^{-1}(1\pm\tfrac{1}{4}r^2)^{-2}\{d\bar{x}^2+d\bar{y}^2+d\bar{z}^2\}.$$

From this we see that $|K|$ determines the scale of length: the two-dimensional analogue is the curvature of the sphere which determines its size. The quantity $K/|K| = \pm1$, if $K \neq 0$, determines the type of geometry, so we have indeed essentially three possibilities here. For ease of notation we define $k = K/|K| = \pm1$ if $K \neq 0$, and $k = 0$ if $K = 0$.

Notice next that the metric (6.4.1) satisfies the condition of isotropy about the origin $r = 0$, since the metric coefficients depend only on r^2. The proof of homogeneity, or equivalently, isotropy about any other point, is not so easy. Indeed a casual glance at the metric gives the impression that $r = 0$ is a privileged centre, except in the case $K = 0$ where we already know the space is homogeneous. The simplest proof of homogeneity is obtained by considering the space from the point of view of an embedding in a bigger space.

To see how this works consider again the two-dimensional sphere. This is defined by the metric

$$d\sigma^2 = d\theta^2 + \sin^2\theta \; d\phi^2, \tag{6.4.2}$$

for which it looks as if $\theta = \pi/2$ is a privileged point. But by the transformation

$$x = \sin\theta \, \cos\phi$$
$$y = \sin\theta \, \sin\phi \tag{6.4.3}$$
$$z = \cos\theta$$

we can manifest the sphere as the surface $x^2+y^2+z^2 = 1$ in three-dimensional Euclidean space. For equations (6.4.3) certainly imply x, y, z satisfy this relation. Then, according to (6.4.3) on this surface we have $dx^2 + dy^2 + dz^2 = d\theta^2 + \sin^2\theta \, d\phi^2 \equiv d\sigma^2$, and the surface must therefore be the sphere defined by (6.4.2), as was to be shown. A rotation in the three-dimensional Euclidean embedding space takes any point of the sphere into any other, and so all points are equivalent and the sphere is homogeneous. This way of looking at the problem avoids the calculation of the explicit effect of a rotation on points of the sphere in terms of the θ, ϕ coordinates, and the demonstration that this leaves (6.4.2) unchanged in form.

For the metric (6.4.1) we make the following transformation:

$$x = \frac{k^{1/2}r}{1+\tfrac{1}{4}kr^2} \sin\theta \, \cos\phi, \qquad y = \frac{k^{1/2}r}{1+\tfrac{1}{4}kr^2} \sin\theta \, \sin\phi,$$

$$z = \frac{k^{1/2}r}{1+\tfrac{1}{4}kr^2} \cos\theta, \qquad w = \frac{1-\tfrac{1}{4}kr^2}{1+\tfrac{1}{4}kr^2}.$$

After a certain amount of tedious but routine algebra we conclude that

$$x^2 + y^2 + z^2 + w^2 = 1,$$

and that

$$dx^2 + dy^2 + dz^2 + dw^2 = (1+\tfrac{1}{4}Kr^2)^{-2}\{dx^2 + dy^2 + dz^2\}.$$

It follows that the geometry described by (6.4.1) can be thought of as a three-dimensional sphere in four-dimensional Euclidean space, and hence that a rotation in the embedding space takes any point of the sphere into any other. This demonstrates the announced homogeneity.

Note that while this description is mathematically correct as an analytic demonstration, it does not provide a correct physical picture of the geometry in all cases. For if $k = -1$, the x, y, z coordinates are imaginary and the metric of the embedding space is Lorentzian, not Euclidean. In this case, therefore, the surface is 'really' a hyperboloid in Minkowski space, and the rotation is 'really' a Lorentz transformation. However, this information is not required for the analytical demonstration of homogeneity.

We now have to show that the metrics (6.4.1) are the only possible homogeneous and isotropic geometries. Note that this is not the same as showing that there are only three types of homogeneous and isotropic three-dimensional surfaces in four-dimensional space since there might be further possibilities which cannot be embedded in four dimensions. In fact, it turns out that there are no such possibilities. We shall show first that a homogeneous and isotropic space must be a space of constant curvature (to be defined below), and then outline the argument that there are only three types of spaces of constant curvature.

Note first that the x, y, z coordinates of (6.4.1) are normal coordinates at the origin, for we can write

$$d\sigma^2 = (1-\tfrac{1}{2}Kr^2 + \ldots)(dx^2 + dy^2 + dz^2).$$

The most general expression for the metric in normal coordinates is

$$d\sigma^2 = (\delta_{kl} + g_{ijkl}\xi^i\xi^j + \dots)(d\xi^k\, d\xi^l),\qquad(6.4.4)$$

with a sum over $i, j, k, l = 1, 2, 3$ understood, as usual. Isotropy about $r = 0$ is imposed by restricting the coefficients g_{ijkl} to be the components of an isotropic (Cartesian) tensor. By definition, this will ensure that the form of the metric is unchanged to this order if the coordinate system is rotated through an arbitrary angle. The most general form for an isotropic fourth-rank tensor (Jeffreys 1956) is

$$\alpha\delta_{ij}\delta_{kl} + \beta(\delta_{ik}\delta_{jl} + \delta_{il}\delta_{kj}) + \gamma(\delta_{ik}\delta_{jl} - \delta_{il}\delta_{kj}),$$

with α, β and γ arbitrary constants. However, we can assume g_{ijkl} to satisfy certain symmetry requirements. In particular, as in §5.14, g_{ijkl} is symmetric in each pair of indices

$$g_{ijkl} = g_{jikl},\qquad g_{ijkl} = g_{ijlk}.$$

It follows from this that $\gamma = 0$.

Now the form of g_{ijkl} can be altered by making a coordinate transformation of third order in x. But, as described in §5.4, the quantity $R_{likj} = (g_{ijkl} - g_{ilkj} - g_{kjil} + g_{klij})$ has a physical significance independent of the choice of further coordinate transformations. It is called the *Riemannian curvature of the space*, evaluated here in normal coordinates and at the origin $r = 0$. We get

$$R_{likj} = 2(\alpha - \beta)(\delta_{ij}\delta_{kl} - \delta_{ik}\delta_{jl}).\qquad(6.4.5)$$

For a general metric this result holds only at the origin of normal coordinates, but in our case it follows from the homogeneity requirement that (6.4.5) must be valid at all points, with $(\alpha - \beta)$ a constant, independent of where we place the origin of the normal coordinate system. Up to a rescaling of the unit of length, we see that at this stage there are just three distinct possibilities, namely, $2(\alpha - \beta) = k = \pm 1, 0$. A space in which (6.4.5) holds at all points with $(\alpha - \beta)$ a constant is said to have constant curvature. The argument has therefore shown that a space which is isotropic about each of its points is a space of constant curvature.

It might be thought that similar arguments could be applied to the higher-order terms in (6.4.4) to generate an infinite number of different combinations of possibilities. However, once the curvature is known *everywhere*, not just at a point, then in general all the higher terms are also determined. For the curvature is related to derivatives of the metric up to second order (equation 5.14.12). Once the curvature is known everywhere, this relation gives a set of differential equations from which the metric can be determined (see e.g. Weinberg 1972). The result of this calculation is that there are indeed no further possibilities and there are just three spaces of constant curvature.

We now have to use this result to construct the most general four-dimensional space–time geometry having homogeneous and isotropic spatial sections of constant cosmic time. This is straightforward: the metric cannot contain terms of the form dx dt since these would change sign under $x \to -x$, thereby violating isotropy. We therefore have

$$d\tau^2 = f^2(T)\, dT^2 - |K(T)|^{-1}[1+(k/4)\,r^2]^{-2}\,(dx^2+dy^2+dz^2),$$

where K is now an arbitrary function of the time coordinate, T, and f is an arbitrary function. A transformation of time coordinate

$$T \to t = \int^T f(T)\, dT,$$

and a relabelling of symbols, takes the metric to the final form

$$d\tau^2 = dt^2 - a^2(t)[1+(k/4)\,r^2]^{-2}(dx^2+dy^2+dz^2) \qquad (6.4.6)$$

where $k = 0, \pm 1$. The function $a(t)$ is to be determined by the Einstein equations. As discussed in connection with the metric (5.11.1), to obtain a time coordinate measured in seconds, we use t/c instead of t.

This result was derived independently by Robertson and Walker, and (6.4.6) is often called the *Robertson–Walker metric*. That such a metric was a possible solution of Einstein's equations, was found much earlier by Friedmann, and the properties of such metrics in the general case $\Lambda \neq 0$ were investigated by Lemaitre. Consequently the metric (6.4.6) is sometimes referred to by combinations and permutations of these names.

The form (6.4.6) is the most natural one from the point of view of normal coordinates, but it is not necessarily the most useful for general calculations, and other coordinates are often employed. Thus the transformation

$$R = \frac{r}{1+\tfrac{1}{4}Kr^2}, \quad x = r \sin\theta \cos\phi, \quad y = r \sin\theta \sin\phi, \quad z = r \cos\theta$$

leads to

$$d\tau^2 = dt^2 - a^2(t)\left\{ \frac{dR^2}{1-kR^2} + R^2(d\theta^2 + \sin^2\theta\, d\phi^2) \right\}. \qquad (6.4.7)$$

From this we see that the area of a sphere of 'radius' R is $4\pi R^2 a^2(t)$, so R has a physical significance as a measure of area. Note that $Ra(t)$ is not the radius measured by a rigid ruler: the distance between points with coordinates (r, R, θ, ϕ) and $(t, R+dR, \theta, \phi)$ is $d\sigma = a(t)(1-kR^2)^{-1/2}\, dR$, and not just $a(t)\, dR$. This is why we write 'radius' in inverted commas.

We shall show below that freely falling observers follow $\mathbf{x}$ = constant lines. The ruler measured distance from $R = 0$ to $R = R_1$, as laid out at time t for a hypothetical set of such observers is

$$a(t) \int_0^{R_1} (1-kR^2)^{-1/2}\, dR = a(t)\, \frac{\sin^{-1}(\sqrt{k}R_1)}{\sqrt{k}}.$$

130

which is $\geq a(t)R_1$ according as $k \lessgtr 0$, and equals $a(t)R_1$ for $k = 0$. This expresses the difference between hyperbolic ($k < 0$) and spherical ($k > 0$) geometry.

Alternatively, we can use coordinates in which the radial-looking coordinate is really proportional to a radius, χ, but the area is not $4\pi\chi^2a^2(t)$. To obtain these put

$$d\chi = \frac{dR}{(1-kR^2)^{1/2}}.$$

By integrating we get

$$R = \frac{\sin\sqrt{k}\chi}{\sqrt{k}},$$

and so

$$d\tau^2 = dt^2 - a^2(t)\left\{d\chi^2 + \frac{\sin^2\sqrt{k}\chi}{k}(d\theta^2 + \sin^2\theta\ d\phi^2)\right\}. \qquad (6.4.8)$$

In the case $k = +1$ it is easy to see by comparison with the discussion of the Einstein universe in §6.2 that the spatial geometry of the Robertson-Walker metric is spherical. The forms (6.4.6), (6.4.7) and (6.4.8) are those most commonly encountered in the literature.

6.5. Interpretation of the Metric

The metric controls the paths of freely falling test particles and rays of light, and these in turn determine the meaning of the metric. We can find the equations of geodesics either from the Euler-Lagrange equations (§5.14) or, as in §5.11, by transforming to normal coordinates (λ, ξ) at an arbitrary point (λ_P, ξ_P). The transformation

$$t - t_P = \lambda - \lambda_P - \tfrac{1}{2}(\dot{a}/a)_P(\xi - \xi_P)^2$$

$$x - x_P = \frac{1}{a_P}(\xi - \xi_P) - \tfrac{1}{2}(\dot{a}/a^2)_P(\lambda - \lambda_P)(\xi - \xi_P) + \frac{k}{4a_P^3}\xi_P(\xi - \xi_P)^2$$

$$+ \frac{k}{4a_P^3}[\xi_P \cdot (\xi - \xi_P)](\xi - \xi_P)$$

reduces the metric (6.4.6) to normal form at P. The geodesic equations

$$\left.\frac{d^2\lambda}{d\tau^2}\right|_P = 0; \qquad \left.\frac{d^2x}{d\tau^2}\right|_P = 0$$

become

$$t'' + \frac{a\dot{a}}{(1+\tfrac{1}{4}kr^2)^2}x' \cdot x' = 0$$

$$x'' + (2\dot{a}/a)x't' - \tfrac{1}{2}kx'^2x - kx \cdot x'x = 0$$

131

where the prime denotes differentiation with respect to τ. A solution is

$$\mathbf{x} = \text{constant} \qquad t = \tau, \qquad (6.5.1)$$

which is compatible with (6.4.6).

The solution (6.5.1) also describes the path of fundamental observers who are at rest in the homogeneous spatial sections of constant cosmic time t. It follows that the fundamental observers are freely falling. We should like to interpret (6.5.1) also as the trajectories of clusters of galaxies in an expanding universe with scale factor $a(t)$. There are two questions that have to be investigated before we can do this. First, we do not measure velocities of receding galaxies: we measure their redshifts. Thus we have to check that the theory reproduces the redshift effect. This we shall do in a moment. Secondly, clusters of galaxies are not test particles moving in a prearranged geometry. The geometry of the Universe is itself determined by the clusters of galaxies, and any one cluster moves in the gravitational field of all the others. Thus the metric must be consistently generated by the clusters, and their assumed motions, through Einstein's field equations. This will be taken up in the next section.

To evaluate the spectral shift of light from distant galaxies in the Robertson–Walker geometry we use the Equivalence Principle. This tell us that over short time intervals in freely falling frames the light waves behave as in special relativity. Therefore, at an arbitrary point on the ray we introduce normal coordinates (λ, ξ). To a freely falling observer at a neighbouring point the frequency is shifted by the special relativistic Doppler effect. In §5.12 it was shown that neighbouring fundamental observers have a relative velocity $\dot{a}/a\,\delta\xi$ in the $k = 0$ Robertson–Walker metric (5.11.1). An analogous argument, using the corresponding coordinate transformation to be discussed in §6.6, shows that this result is also true in the $k = \pm 1$ cases. Therefore,

$$-\left(\frac{\delta\nu}{\nu}\right)_P = \frac{1}{c}\left\{\frac{1}{a}\frac{da}{d(t/c)}\,|d\xi|\right\}_P = \{(\dot{a}/a)|\delta\xi|\}_P$$

to first order in small quantities. For a light ray in special relativity we have $|\delta\xi| = \delta\lambda$, and so

$$-\left(\frac{\delta\nu}{\nu}\right)_P = \left\{\left(\frac{\dot{a}}{a}\right)\delta\lambda\right\}_P = \left\{\frac{1}{a}\frac{da}{d\lambda}\,\delta\lambda\right\}_P$$

or

$$-\left(\frac{d\log\nu}{d\lambda}\right)_P = \left(\frac{d\log a}{d\lambda}\right)_P.$$

Since the coordinates are normal at P only, this equation holds only at P, and we cannot integrate it as it stands. However, we can transform to $(t, \mathbf{x})$ coordinates

132

to obtain, at P,

$$-\frac{d \log \nu}{dt} = \frac{d \log a}{dt} .$$

But now, since P is an arbitrary point in the space–time and has no special significance for the coordinates $(t, \mathbf{x})$, this equation holds everywhere. We can therefore integrate it to obtain the fundamental result

$$1 + z = \frac{\nu}{\nu_0} = \frac{a_0}{a} , \qquad (6.5.2)$$

where $a_0 = a(t_0)$. Consequently, if galaxies are taken to follow geodesics, $\mathbf{x} = $ constant, we shall observe redshifts given by (6.5.2). For small z, and hence small $(t - t_0)$, we have

$$1 + z \sim 1 + \frac{\dot{a}}{a} \, \delta t$$

or

$$z \sim \frac{1}{a} \frac{da}{d(t/c)} \frac{d}{c} ,$$

where $\delta t = d/c$, which is the Hubble law. Therefore, we can identify $a(t)$ with the scale factor $R(t)$ subject to the discussion of the next section.

Since this result is so important we shall give an alternative derivation which avoids the use of normal coordinates. Consider the emission of successive wave-crests of light by a galaxy G at time t_e, $t_e + \delta t_e$ (figure 6.3). These are received by an observer O, at the origin of the $(t, \mathbf{x})$ coordinates at times t_0, $t_0 + \delta t_0$. The object of the calculation is to relate the time intervals δt_e, δt_0, at emission and reception and hence the frequencies of the light.

Along a light ray we have $d\tau^2 = 0$. Either by computation of null geodesics, or more simply by a symmetry argument, we know that a light ray initially in the x, t plane will remain in this plane. We can therefore put $dy = 0$, $dz = 0$ along a ray. Hence, $d\tau^2 = 0$ in (6.4.6) implies

$$\frac{dt}{a(t)} = \frac{-dx}{1 + kx^2/4}$$

for a light ray moving towards the origin (x decreasing as t increases). The 'comoving separation' of O and G is

$$R \equiv -\int_{x_G}^{0} \frac{dx}{(1 + kx^2/4)} = +\int_{t_e}^{t_0} \frac{dt}{a(t)} . \qquad (6.5.3)$$

Since fundamental observers travel along $\mathbf{x} = $ constant curves, R is independent of

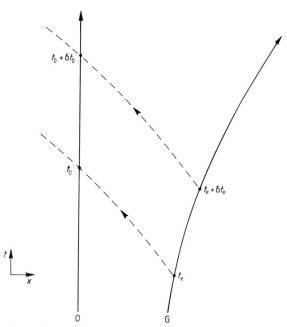

Figure 6.3. The emission of successive wave crests of light by galaxy G, and their reception at galaxy O are labelled by the values of the cosmic time, t.

time. Therefore

$$R = \int_{t_e + \delta t_e}^{t_0 + \delta t_0} \frac{dt}{a(t)}.$$ (6.5.4)

From (6.5.3) and (6.5.4) it follows that

$$\int_{t_0}^{t_0 + \delta t_0} \frac{dt}{a(t)} - \int_{t_e}^{t_e + \delta t_e} \frac{dt}{a(t)} = 0$$

and, consequently,

$$\frac{\delta t_e}{a(t_e)} = \frac{\delta t_0}{a(t_0)}.$$

Since fundamental observers measure proper times, we find for the emitted frequency ν_e, and observed frequency ν_0 the relation

$$1 + z = \frac{\nu_e}{\nu_0} = \frac{\delta t_0}{\delta t_e} = \frac{a(t_0)}{a(t_e)},$$

which is the same as (6.5.2).

134

We can now introduce the Hubble parameter, H, and the deceleration parameter, q, in direct analogy with the elementary discussion of Chapter 2. We put

$$H = c\,\frac{\dot{a}(t)}{a(t)},$$

$$q = -\frac{\ddot{a}(t)\,a(t)}{\dot{a}^2(t)},$$

and again use H_0 and q_0 to denote the present values of these parameters. The extra factor of c in the Hubble parameter arises from our choice of units for the coordinate t, and disappears if we replace t by a coordinate measured in seconds. Einstein's equations will be shown to determine higher derivatives of $a(t)$, so these parameters, together with k, define a cosmological model.

6.6. The Field Equations

By insisting that the galaxies determine the gravitational fields in which they move, we shall find constraints on the Robertson–Walker metrics which restrict the possible forms for $a(t)$. This is scarcely surprising, since we should expect that the more galaxies we have the greater the gravity opposing the expansion, and the greater the slowing down of that expansion. What does come as a complete surprise, however, is the relation of the geometry of space to the density of matter; beyond a certain density we shall find that space closes up on itself and the models must be 'spherical', having $k = +1$.

To obtain this crucial result we shall need the field equations as they apply to the general Robertson–Walker metric. These are provided unambiguously in general relativity. We shall simply try to illuminate the result by constructing a form, valid for general k, that reduces to our previous result for the case $k = 0$.

First we need the transformation to coordinates suitable for the discussion of geodesic deviation analogous to that in §5.12. Write $\psi = [1 + (k/4)\,r^2]$ for simplicity. The transformation

$$t = \lambda - \tfrac{1}{2}\dot{a}/a\,\xi^2$$

$$\mathbf{x} = \frac{1}{a}\,\boldsymbol{\xi} + \frac{1}{2a^2}\,(\xi^2 \nabla\psi + 2\boldsymbol{\xi}\,\boldsymbol{\xi}\cdot\nabla\psi)$$

yields

$$dt = d\lambda\,\left(1 - \tfrac{1}{2}\frac{\ddot{a}}{a}\,\xi^2 + \tfrac{1}{2}\frac{\dot{a}^2}{a^2}\,\xi^2\right) - \frac{\dot{a}}{a}\,\boldsymbol{\xi}\cdot d\boldsymbol{\xi}$$

$$d\mathbf{x} = -\frac{\dot{a}}{a^2}\,\boldsymbol{\xi}\,d\lambda + \frac{1}{a}\,d\boldsymbol{\xi} + \tfrac{3}{4}k\,\frac{\xi^2}{a^3}\,d\boldsymbol{\xi} + \tfrac{3}{2}\frac{k}{a^3}\,\boldsymbol{\xi}\,(\boldsymbol{\xi}\cdot d\boldsymbol{\xi})$$

135

to second order, and hence, as required,

$$d\tau^2 = d\lambda^2 \left(1 - \frac{\ddot{a}}{a}\xi^2\right) - d\xi^2 \left(1 - \xi^2(\dot{a}/a)^2 + \frac{k}{a^2}\xi^2\right) + [(\dot{a}/a)^2 - 3k/a^2](\xi \cdot d\xi)^2$$

by substitution in (6.4.6). The line $\xi = 0$ is the world-line of a freely falling observer. In (λ, ξ) coordinates, physical processes are described to first order, near $\xi = 0$, according to special relativity.

An observer near $\xi = 0$ falls along a geodesic given by

$$\frac{d^2\xi}{d\tau^2} - \frac{\ddot{a}}{a}\xi - 2\left[\left(\frac{\dot{a}}{a}\right)^2 - k/a^2\right]\left(\xi \cdot \frac{d\xi}{d\tau}\right)\frac{d\xi}{d\tau} + 2k/a^2\left(\frac{d\xi}{d\tau} \cdot \frac{d\xi}{d\tau}\right)\xi = 0 \qquad (6.6.1)$$

$$\lambda = \tau$$

to first order in $|\xi|$. These equations are obtained, as in §5.12, either by transforming to normal coordinates at a point near $\xi = 0$, by transformation of the geodesic equations, or, most easily, from the general theory of §5.14. If $k = 0$, we regain equation (5.12.6).

Not all the terms in (6.6.1) have a physical significance, since the form of the equation can be altered by a further coordinate transformation of third order in $|\xi|$. Let us write (6.6.1) as

$$\xi''^i + S^i_{0j0}\xi^j + S^i_{jkl}\xi'^j\xi'^k\xi'^l = 0,$$

where

$$S^i_{0j0} = -\delta^i_j \ddot{a}/a$$

$$S^i_{jkl} = (\delta^i_j\delta_{kl} + \delta^i_l\delta_{kj})[k/a^2 - (\dot{a}/a)^2] + 2k/a^2\,\delta^i_k\delta_{jl}$$

are ordinary Cartesian tensors in the Euclidean space near $\xi = 0$. The quantities of physical significance turn out to be $S^i_{jkl} - S^i_{jlk}$ and S^i_{0j0}, and we define

$$R^i_{0j0} = -R^i_{00j} = S^i_{0j0} = -\delta^i_j\,\ddot{a}/a$$

$$R^i_{jkl} = S^i_{jkl} - S^i_{jlk} = -(\delta^i_l\delta_{kj} - \delta^i_k\delta_{lj})[(\dot{a}/a)^2 + k/a^2].$$

The quantities so obtained are essentially the components of the Riemannian curvature defined in §5.14. They may be obtained by comparison of the particular metric form here with (5.14.10) and using (5.14.11).

We can now reconstruct the field equations of §5.13 for $k = 0$ in a systematic way. There are several intermediate steps. First we compute components of the so-called Ricci tensor:

$$R^0_0 = R^i_{0i0} = -3\ddot{a}/a$$

$$R^i_k = R^i_{0k0} - \delta^{jl}R^i_{jkl} = -\delta^i_k[\ddot{a}/a + 2(\dot{a}/a)^2 + 2k/a^2],$$

and of the Ricci scalar

$$R = R^0_0 + R^i_i = -6[\ddot{a}/a + (\dot{a}/a)^2 + k/a^2].$$

Next we construct the Einstein tensor

$$G_\nu^\mu = R_\nu^\mu - \tfrac{1}{2}\delta_\nu^\mu R.$$

This is the simplest geometrical quantity constructed out of the curvature that can be set equal to the quantity which represents the flow of energy and momentum in special relativity (the energy-momentum tensor). In our particular case the material content of the space-time model is described by the energy density and pressure and the field equations simplify to

$$G_0^0 = \frac{8\pi G}{c^2}\mu \tag{6.6.2}$$

$$G_1^1 = G_2^2 = G_3^3 = -\frac{8\pi G}{c^4}p. \tag{6.6.3}$$

Written out explicitly, these are

$$3\left(\frac{\dot{a}^2}{a^2} + \frac{k}{a^2}\right) = \frac{8\pi G}{c^2}\mu \tag{6.6.4}$$

$$2\frac{\ddot{a}}{a} + \frac{\dot{a}^2}{a^2} + \frac{k}{a^2} = -\frac{8\pi G}{c^4}p. \tag{6.6.5}$$

Now for $k = 0$, (6.6.4) and (6.6.5) reduce to equations equivalent to the field equations (5.13.5) and (5.13.6). We can therefore adopt them as the required field equations in the general case of $k = 0$. Note again that all of the ambiguities in this procedure are completely resolved in the full theory.

An equivalent, and often useful way of writing (6.6.5) is

$$\frac{\ddot{a}}{a} = -\frac{4\pi G}{3c^2}(\mu + 3p/c^2) \tag{6.6.6}$$

which is obtained from (6.6.5) on using (6.6.4). The energy conservation equation is, from (5.13.5),

$$\frac{d}{dt}(\mu a^3) + \frac{p}{c^2}\frac{d(a^3)}{dt} = 0. \tag{6.6.7}$$

That this is compatible with the field equations is readily verified by substituting for μ and p from (6.6.4) and (6.6.5) in the left-hand side of (6.6.7). Together (6.6.6) and (6.6.7) are equivalent to (6.6.4) and (6.6.5), so either pair may serve as the field equations.

If the cosmological constant Λ is non-zero, these equations require modification. In particular, (6.6.4) becomes

$$3\left(\frac{\dot{a}^2}{a^2} + \frac{k}{a^2}\right) - \Lambda = \frac{8\pi G}{c^2}\mu, \tag{[(6.6.4)']}$$

137

while (6.6.5) is replaced by

$$2\frac{\ddot{a}}{a} + \frac{\dot{a}^2}{a^2} + \frac{k}{a^2} - \Lambda = -\frac{8\pi G}{c^4}p. \qquad [(6.6.5)']$$

Equations (6.6.4) and (6.6.5) ensure that in the Robertson-Walker cosmologies the matter generates the gravitational field in which it moves. Since we have two equations for the three unknown functions $[\mu(t), p(t), a(t)]$, the equations would appear to represent no restriction at all on $a(t)$. However, for all normal matter, p and μ are not independent but must be related by some equation of state, the form of which depends on the type of matter under consideration. In the next two sections we shall consider two extreme cases, which approximate to the conditions prevailing at different stages of the evolving universe.

Before leaving the field equations, however, we may note an intriguing relation to Newtonian dynamics. Imagine a spherically symmetric expanding ball of gas with fixed centre at $r = 0$. Consider a particle moving along with the fluid at radius $r(t)$, at time t, so that inside $r(t)$ the amount of matter is constant. The matter outside $r(t)$ has no gravitational effect on this particle and may be neglected. Equating the force per unit mass to the acceleration we find the equation of motion of the particle:

$$\ddot{r} = -\frac{GM(r)}{r^2}, \qquad (6.6.8)$$

where $M(r)$ is the mass inside $r(t)$. Now write $r(t) = R(t)\,r_0$, assuming the expansion to be uniform as in §2.3. In addition we put

$$M(r) = \frac{4\pi}{3}r^3\rho.$$

Then (6.6.8) yields

$$\frac{\ddot{R}}{R} = -\frac{4\pi G}{3}\rho \qquad (6.6.9)$$

which, apart from the unit of time, is the same as equation (5.13.1).

Equation (6.6.9) can also be compared with the general relativistic equation (6.6.6). We see that if we identify $a(t)$ with $R(t)$ these equations are the same in the case $p = 0$; for in this case there is no internal random motion, and so $\mu = \rho$.

Now in the case $p \neq 0$ we do not expect a Newtonian argument to reproduce the general relativistic result, since in relativity the motion of particles itself produces a contribution to gravity over and above its effect through the energy density μ. There is a direct analogy here with electromagnetic theory, where charges in motion produce magnetic fields in addition to their electrostatic effects. However, the real point is that one does not really expect to reproduce the general relativistic result by a Newtonian argument in any case! So the question arises as to

138

whether the similarity is merely an accident. The answer is a qualified no. Certainly a parallel development of Newtonian gravity and general relativity can be made in which, apart from the replacement $\rho \rightarrow \mu + 3p$, some of the field equations come out looking the same (Ellis 1971). In this particular case the high degree of symmetry means that only the one field equation (6.6.6) and the conservation law (6.6.7) survive; the remainder, which in general would provide the difference between relativity and Newtonian gravity, reduce here to identities (i.e. to 0 = 0).

Nevertheless, this parallel cannot be drawn between relativity and the *standard* form of Newtonian theory. In particular, all points of our model universe are supposed to be equivalent, yet in the Newtonian analysis we appear to require a preferred centre in order to define the origin of an inertial reference frame. All other galaxies are accelerated relative to this centre, so cannot be the origins of inertial frames, and hence cannot be equivalent. This is circumvented by reformulating Newtonian theory under the aegis of the Equivalence Principle. Instead of formulating an equation of motion like (6.6.8) in inertial frames, this is taken to apply in freely falling frames, hence for any fundamental observer, and is reinterpreted as an equation governing the deviation of free-falls.

A second difficulty is that (6.6.9) is an equation of motion of a particle, whereas (6.6.6) is a field equation, analogous to Poisson's equation. How then can it be anything but an accident that the two are similar? The answer is related to the fact that in general relativity the field equations and the equations of motion of particles are not independent: the former imply the latter. In fact, equation (6.6.9) *can* be derived directly in terms of the relative motion of freely falling particles. We conclude that the argument leading to (6.6.9) is *not* a correct derivation of the field equation, even in the case $p = 0$, but that there are deep reasons why the answer happens to come out right.

6.7. The Dust Universe

The results of Chapters 3 and 4 show that at the present time the energy density in radiation in the Universe is much less than that in matter. We know also from Chapter 2 that the random motions of galaxies and clusters of galaxies are small, in the sense that the kinetic energy of random motion, excluding, that is, the Hubble expansion velocity, is much less than the rest mass–energy density. Consequently we have $\mu \approx \rho$, and an approximate equation of state for the cosmic gas of galaxy clusters is

$$p = 0.$$

Cosmological tradition has it that such a material is called 'dust' for reasons which, if they ever existed, have not been transmitted to later generations of cosmologists. In the dust model of the Universe it is assumed that $p = 0$ at *all* times. It is therefore an inappropriate model at early times.

Let us consider first the Einstein model of a static universe with $\Lambda \neq 0$. Putting $p = 0, \dot{a} = 0, \ddot{a} = 0$ in $(6.6.5)'$ gives

$$\Lambda = k/a_E^2, \qquad (6.7.1)$$

where a_E is the constant value of a. From $(6.6.4)'$ we then obtain

$$\Lambda = \frac{4\pi G}{c^2} \rho_E, \qquad (6.7.2)$$

where ρ_E is the constant value of the density. Since $\rho_E > 0$, (6.7.2) implies $\Lambda > 0$ and then, from (6.7.1), $k > 0$. Therefore the only static model with $\Lambda \neq 0$ has closed, spherical spatial sections and we obtain the metric (6.2.1).

It is easy to show that a static model with $\Lambda = 0$ is not possible for a reasonable equation of state. For if $\ddot{a} = 0$ in (6.6.6) we must have $\mu + 3p/c^2 = 0$. Since, obviously, $\mu > 0$ this can only hold if we allow unphysical negative pressures. We therefore return to the dust-filled universe and consider time-dependent models. For simplicity from now on we shall take $\Lambda = 0$.

From the conservation equation (6.6.7), or from the discussion in §5.13, we obtain, in the case $p = 0$,

$$\rho a^3 = \text{constant} = \rho_0 a_0^3. \qquad (6.7.3)$$

The field equation (6.6.4) implies

$$\tfrac{1}{2}\dot{a}^2 - \frac{G}{c^2}\left(\frac{4\pi}{3} \rho_0 a_0^3\right)\bigg/ a^2 = -k/2. \qquad (6.7.4)$$

We can interpret this as an energy equation: in Newtonian language $\tfrac{1}{2}\dot{a}^2$ is the kinetic energy per unit mass of a particle with velocity $\dot{a}$; the second term is the potential energy per unit mass due to the sphere of matter interior to the particle at radius a. The equation as a whole then states that kinetic energy plus potential energy is conserved.

Before we proceed to integrate (6.7.4) to determine $a(t)$, we can derive some useful results. Using the definition of H, equation (6.7.4) gives

$$c^2 k = a^2\left(\frac{8\pi G}{3} \rho - H^2\right). \qquad (6.7.5)$$

By considering this relation at the present time it follows that $k > 0$ implies

$$\rho_0 > \frac{3H_0^2}{8\pi G},$$

and $k < 0$ if

$$\rho_0 < \frac{3H_0^2}{8\pi G}.$$

The density

$$\rho_c = \frac{3H_0^2}{8\pi G}$$

is therefore a 'critical' density at the present time, separating those models with closed, spherical spatial sections of constant cosmic time from those with infinite, hyperbolic spatial sections. The critical case itself has flat spatial sections. The field equations lead immediately, therefore, to the conclusion that (at least for $\Lambda = 0$) a knowledge of the Hubble constant, and the matter density determines the type of geometry of space.

Using $H_0 = 100\,h$ km s^{-1} Mpc, we get

$$\rho_c = 1.9 \times 10^{-26} h^2 \text{ kg m}^{-3}.$$

If all the matter in the Universe were in galaxies we should have $\rho_0 < \rho_c$ (§4.6), and we could conclude that the Universe is 'open', i.e. that space extends to infinity. This is part of the promised importance of the present matter density (§2.8), for it determines the type of spatial geometry. As we shall see presently, it also determines the type of future.

Equation (6.6.6), with $p = 0$, $\mu = \rho$, can be written as

$$q_0 = \frac{4\pi G\rho_0}{3H_0^2} = \frac{1}{2}\frac{\rho_0}{\rho_c} \tag{6.7.6}$$

at the present epoch. Thus a further consequence of the field equations is $q_0 \gtrless \frac{1}{2}$ if $\rho_0 \gtrless \rho_c$: a measurement of q_0 would yield the matter density directly, and hence the geometry of space. At present q_0 cannot be measured with sufficient accuracy to distinguish the possible cases (Chapter 7). Note that since k cannot change sign, it follows from (6.7.5) and (6.7.6) that, unless $q = \frac{1}{2}$, we have either $q > \frac{1}{2}$ for all time or $q < \frac{1}{2}$ for all time.

There are now three cases to be discussed separately.

(i) *The case k = 0* (the Einstein–de Sitter model). Equation (6.7.3) is easily integrated to give

$$a(t) = (6\pi G\rho_0 a_0^3)^{1/3}(t/c)^{2/3}, \tag{6.7.7}$$

in which the constant of integration has been chosen such that $a(0) = 0$. Note that whatever the constant of integration, $a(t)$ must become zero at some finite time in the past, and it makes no difference to the physics that we use this convenient convention.

Loosely speaking, we may say that at $t = 0$ we cannot make a transformation to normal coordinates, since the coefficients in the transformation and the second-order terms in the metric become infinite. The idea of the Equivalence Principle therefore breaks down at $t = 0$, and we say we have a 'singularity'. Certainly at this point the density of matter will be infinite, and we shall not be able to do physics.

In addition, the redshift, $z = a_0/a - 1$, becomes infinite, so we shall not receive any radiation from this time. Thus we cannot discuss what happens at $t = 0$, or before, and this model begins with an 'unknown' big bang. A slightly more precise discussion is given in Chapter 11.

As time increases, $a(t)$ increases monotonically. The galaxies move apart but their velocities of separation, and the redshifts of given galaxies tend to zero. This model is called the *Einstein–de Sitter model*. We can calculate the present age of our Universe in terms of measured quantities to replace the estimate $t_0 \sim H_0^{-1}$ in §2.6. We have

$$H_0 = c \left(\frac{\dot{a}}{a} \right)_0 = \frac{2}{3} \frac{1}{(t_0/c)}$$

from (6.7.6), or $t_0/c = \frac{2}{3} H_0^{-1} = 7 \times 10^9 h^{-1}$ years. Of course, while this is exact for the Einstein–de Sitter model, it is at best only an approximation for the real Universe, since we shall find that the dust model is not valid at early times.

In general, if

$$\frac{4\pi G}{3c^2} \rho_0 a_0^3/a > |k|,$$

i.e. for sufficiently small a, we can neglect k in equation (6.7.4). Thus, at sufficiently early times, all of these dust models behave like the Einstein–de Sitter model no matter what the value of k. In particular, they all start from a big bang.

(ii) *The case $k = +1$*. From (6.7.4) we find $\dot{a} = 0$ if

$$a_{\max} = \frac{8\pi G}{3c^2} \rho_0 a_0^3.$$

This universe expands from $a = 0$ to the maximum radius, $a_{\max}$, beyond which it must contract to $a = 0$ again.

We can calculate the age of the Universe at the time of maximum expansion in terms of quantities that can in principle be measured at any time. To do this we write

$$t = \int_0^t dt = \int_0^{a(t)} \frac{dt}{da} \, da. \tag{6.7.8}$$

We can substitute for $dt/da = 1/\dot{a}$ from (6.7.4), which here reads

$$\dot{a} = \left(\frac{8\pi G \rho_0 a_0^3}{3ac^2} - 1 \right)^{1/2}. \tag{6.7.9}$$

By (6.7.6) we can eliminate ρ_0 in favour of q_0; evaluating (6.7.9) at t_0 and rearranging yields

$$a_0 = (2q_0 - 1)^{-1/2} c H_0^{-1}. \tag{6.7.10}$$

This is an important relation since it tells us that a_0 can be expressed in terms of measurable quantities. Note that it does not contradict our previous assertion that only ratios of scale factors can be measured: at this stage we have lost the freedom to rescale units, and hence $a(t)$, since we have chosen them to make $k = +1$.

Using (6.7.10) in (6.7.9) now gives us $\dot{a}$ as a function of a and observable quantities. Substitution in (6.7.8) then gives an integral expression for the age of the Universe:

$$t = \int_0^{a(t)} da \{2q_0(2q_0 - 1)^{-3/2} c H_0^{-1} a^{-1} - 1\}^{-1/2}.$$

The change of variable

$$\sin^2\theta = \frac{(2q_0 - 1)^{3/2} H_0 a}{2q_0 c} \tag{6.7.11}$$

reduces this to

$$[2q_0(2q_0 - 1)^{-3/2} c H_0^{-1}]^{-1} t = 2 \int_0^{\theta(t)} \sin^2\theta = \theta - \sin\theta \cos\theta. \tag{6.7.12}$$

At the point of maximum expansion, $\dot{a} = 0$, from (6.7.9), (6.7.6), (6.7.10) and (6.7.11), we find $\theta = \pi/2$, so the age is

$$t_{max}/c = \pi q_0 (2q_0 - 1)^{-3/2} H_0^{-1}. \tag{6.7.13}$$

The recontraction phase is just the reverse of the expansion phase, since it is obtained by changing the sign of $\dot{a}$, so the total lifetime of this model is $T = 2t_{max}$. Note that t_{max} and T are expressed here in terms of quantities measured at t_0, which is an arbitrary time. We should therefore check that the expression (6.7.13) is independent of t_0. It is a simple matter to show from the field equation (6.6.4), using (6.7.6) and the definition of H, that $(2q - 1)^{3/2} H q^{-1} = $ constant.

We can evaluate also the present age; here $a = a_0$ so $\sin^2\theta = (2q_0 - 1)/2q_0$. Therefore

$$t_0/c = 2q_0(2q_0 - 1)^{-3/2} H_0^{-1} \sin^{-1}(1 - 1/2q_0)^{1/2} - H_0^{-1}(2q_0 - 1)^{-1}.$$

This can be checked by taking the limit $q_0 \to \frac{1}{2}$, in which case we should regain the Einstein–de Sitter model. Expanding about $q_0 = \frac{1}{2}$ by Taylor series (or looking up the first two terms of the expansion of $\sin^{-1}x$) indeed gives $t_0/c = \frac{2}{3} H_0^{-1}$.

For $q_0 > \frac{1}{2}$ we have $t_0/c < \frac{2}{3} H_0^{-1}$; for large q_0 the age of the dust Universe is less than that of the stars it is supposed to contain. Indeed, if $h = 1$ even the Einstein–de Sitter model is ruled out by this consideration. This conclusion is not altered by the inclusion of radiation in the early Universe, since the dust phase covers most of its lifetime.

Note that the expressions for the ages derived here depend only on the parameters q and H. This will be true of observable relations to be derived later in

Chapter 7. No further parameters, involving higher derivatives of the scale factor, are required. This, of course, is a consequence of the field equations by which the higher derivatives are implicitly related to $a, \dot{a}$, and $\ddot{a}$ only.

More important than the precise form of t_{max} is the conclusion that if $q > \frac{1}{2}$, the universe must recollapse. Thus the density of matter not only determines the spatial geometry but also the future fate of the model. This is easily interpreted in Newtonian terms. For if equation (6.7.4) is regarded as an energy equation, then $k = 1$ implies that the total energy is negative; hence the system is bound and cannot expand to infinity. For a sufficient density of matter there is enough gravitational attraction to overcome the initial expansion and induce collapse. At $t = T$ we again get a singularity beyond which the theory is powerless to predict.

(iii) *The case $k = -1$.* In this model there is no radius, $a(t)$, for which $\dot{a} = 0$: for all $a(t)$ we have $\dot{a} > 0$ and the universe expands forever. Near $t = 0$ we can neglect k and again $a(t) \propto t^{2/3}$. For large $a(t)$ we can neglect the potential energy term to get $\dot{a} \sim 1$ or $a(t) \propto t + \text{constant}$. This differs from the $k = 0$ model, where $a(t) \sim t^{2/3}$ as $t \to \infty$, since we have $\dot{a}$ finite at infinity. This behaviour accords with the energy interpretation of equation (6.7.4); $k = -1$ means positive total energy and an unbound system.

For the $k = -1$ model a calculation similar to the $k = +1$ case gives a present age of

$$t_0/c = 2q_0(1-2q_0)^{-3/2}H_0^{-1}\sinh^{-1}(1/2q_0-1)^{1/2}+H_0^{-1}(1-2q_0)^{-1}.$$

For $q_0 = 0.1$ and $H_0 = 100\,\text{km}\,\text{s}^{-1}\text{Mpc}^{-1}$, this is $t_0/c \approx 1.8 \times 10^{10}$ years. Note that the present age in the case $k = -1$ is larger than that for $k = +1$, so there is no problem with stellar evolutionary ages.

All of the interesting properties of these models can be obtained without explicit integration of the field equations. Nevertheless, for completeness, we can describe how the equation can be integrated to give $a(t)$. For $k = 0$ this has already been done (equation 6.7.7).

For $k = +1$, we have from (6.7.12), putting $2\theta = \phi$,

$$t/c = \frac{q_0}{(2q_0-1)^{3/2}H_0}(\phi - \sin \phi). \qquad (6.7.14)$$

From the definition of θ (equation 6.7.11),

$$a(t) = \frac{q_0 c}{(2q_0-1)^{3/2}H_0}(1-\cos \phi).$$

Eliminating ϕ, in principle, gives $a(t)$; in practice we cannot eliminate ϕ but nevertheless these equations give a parametric representation, with parameter ϕ, of a curve in the a–t plane. The curve is, in fact, a cycloid.

144

For the case $k = -1$ we obtain similarly

$$a(t) = \frac{q_0 c}{(1-2q_0)^{3/2}H_0} (\cosh \phi - 1)$$

where   (6.7.15)

$$t/c = \frac{q_0}{(1-2q_0)^{3/2}H_0} (\sinh \phi - \phi).$$

These curves are sketched in figure 6.4. The properties of the solutions can, of course, be obtained directly from these explicit representations.

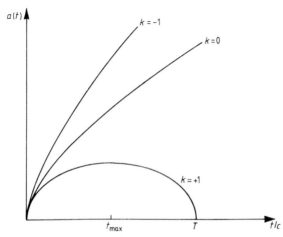

Figure 6.4. Evolution of the scale factor $a(t)$ in the $p = 0$ Robertson-Walker models.

In every Robertson-Walker model there is a unique relation between the observed redshift of a galaxy and the cosmic time at which the light must have been emitted in order to be visible to the observer. Consequently, we can refer to events occurring at redshift z instead of at time t. This relation between t and z will be needed later for the dust models, so we shall derive it in this case. The calculation is similar in other models.

We have, from (6.5.2),

$$1 + z = a(t_0)/a(t),$$ (6.7.16)

and hence

$$\frac{dz}{dt} = -\frac{a_0}{a^2} \dot{a} = -(1+z)\dot{a}/a = -(1+z)\frac{H}{c}.$$ (6.7.17)

145

The object of the exercise is to write this relation in terms of the observable parameters H_0 and q_0. The field equations give us $a(t)$ in terms of ρ and k, so we want to solve for $\dot{a}/a = H$ and eliminate ρ and k in favour of H_0 and q_0. Note that the algebra consists of manipulation of the field equations and the definition of these parameters.

If $p = 0$, (6.6.4) can be written as

$$a^3 H^2 + ka = \frac{8\pi G}{3c^2}\rho a^3 = \frac{8\pi G}{3c^2}\rho_0 a_0^3 = a_0^3 H_0^2 + ka_0 \qquad (6.7.18)$$

on using the conservation equation (6.7.3). This eliminates ρ. The remaining field equation (6.6.5) gives, for $p = 0$,

$$\frac{c^2 k}{a_0^2} = (2q_0 - 1)H_0^2,$$

which is equivalent to (6.7.5). This eliminates k in (6.7.18), which can be solved for H, using (6.7.16) to eliminate a in favour of z. We obtain

$$H = H_0(1+z)(1+2q_0 z)^{1/2},$$

and hence

$$\frac{dz}{dt} = -\frac{H_0}{c}(1+z)^2(1+2q_0 z)^{1/2} \qquad (6.7.19)$$

gives us the required differential equation for $z(t)$. The results of integrating (6.7.19) are plotted in figure 6.5.

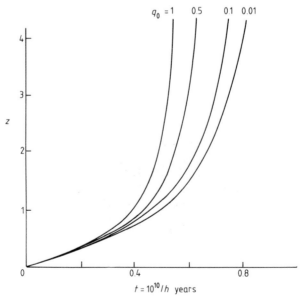

Figure 6.5. $z(t)$ plotted against t from equation (6.7.19). The curves all asymptote $t = 10^{10}h^{-1}$ years as $z \to \infty$.

6.8. The Radiation Universe

At the opposite extreme from a universe filled only with cold matter is a universe filled only with radiation. Such a model will be a suitable approximation to the real world at times when the energy density of radiation, u_r, and hence its gravitational effect, substantially exceeds that of matter. Such a situation must have prevailed at some time in the past, since we know $u_r \propto a^{-4}$, and so increases more rapidly as we go back in time than the matter density $\rho c^2 \propto a^{-3}$. We can calculate the redshift, z_{eq}, at which these are equal from

$$u_r = u_r(t_0) \frac{a_0^4}{a^4} = \rho_0 c^2 \frac{a_0^3}{a^3} = \rho c^2.$$

Therefore

$$1 + z_{eq} = a_0/a = \rho_0 c^2/u_r(t_0) \sim 10^5 q_0.$$

In the Einstein–de Sitter model this corresponds to a time $t_{eq} \sim 10^3$ years, before which the dust model cannot be appropriate.

For a radiation gas we have an equation of state

$$p = \tfrac{1}{3} u_r \tag{6.8.1}$$

and the mass density μ is replaced by u_r/c^2. The field equations (6.6.4) and (6.6.6) become

$$\tfrac{1}{2}\dot{a}^2 - \frac{4\pi G}{3c^4} (u_r a^3)/a = -k/2 \tag{6.8.2}$$

$$\ddot{a} = - \frac{8\pi G}{3c^4} u_r a. \tag{6.8.3}$$

The conservation law (6.6.7) is

$$\dot{u}_r + 4 u_r a^2 \dot{a} = 0, \tag{6.8.4}$$

and this integrates to give $u_r \propto a^{-4}$, which is required for consistency. Equation (6.8.4) is also consistent with the field equations. This is easily confirmed by differentiating (6.8.2) and using $u_r a^4 = \text{constant}$. This yields (6.8.3), so (6.8.2) is the integral of (6.8.3).

We can again interpret (6.8.2) as an energy equation, but the effective mass of radiation inside a given region is no longer constant. By doing work in expanding against gravity, the radiation is redshifted and loses energy. These equations can be treated as in the dust model of §6.7, the only difference being in this potential energy term which has the form constant $\times a^{-2}$ instead of constant $\times a^{-1}$. In fact, this makes explicit integration much easier in the radiation model.

From (6.8.3) we obtain

$$q = \frac{8\pi G}{3c^2} \frac{u_r}{H^2} \tag{6.8.5}$$

which is the analogue of (6.7.6). Using this to eliminate u_r from (6.8.2) we get

$$\frac{c^2 k}{a^2} = H^2(q - 1) \qquad (6.8.6)$$

which is the analogue of (6.7.5). Note that $k \geq 0$ corresponds to $q \geq 1$ in the radiation universe, not $q \geq \frac{1}{2}$ as in the dust model.

There are three cases to be considered separately.

(i) *The case $k = 0$.* From (6.8.2) and (6.8.4) we obtain

$$\frac{a(t)}{a_0} = \left(\frac{32\pi G}{3c^2} u_0\right)^{1/4} \left(\frac{t}{c}\right)^{1/2}, \qquad (6.8.7)$$

which is the equation of a parabola. This is the most important case since all of the models behave in this way at early times; and, of course, the models are only relevant at early times. Equation (6.8.7) gives

$$\frac{a}{a_0} = 10^{-10} \left(\frac{t}{c}\right)^{1/2}$$

where t/c is measured in seconds. But for a black-body spectrum, we know that $T_0/T = a/a_0$. Thus we obtain a relation between the temperature of a universe filled with black-body radiation and the time, a relation which is valid exactly for all time if $q = 1$, and at sufficiently early times in other cases. Explicitly,

$$T = 10^{10}/t^{1/2} \text{ K}. \qquad (6.8.8)$$

At an age of one second, the radiation universe has a temperature of approximately 10^{10} K (see §8.4 for a more accurate discussion).

The relation between temperature and time is an important characteristic of the radiation model; in this respect it is quite different from a matter-filled universe. Both models yield a relation between the energy density and time. The difference arises because the energy density of matter depends on the temperature and on the number density of particles, whereas for black-body radiation both the energy density and the number density of photons are determined by the temperature.

(ii) *The case $k = +1$.* From the conservation equation, $u_r a^4 = \text{constant}$, and equation (6.8.5) we obtain

$$\frac{8\pi G}{3c^2} u_r = \frac{8\pi G}{3c^2} \frac{u_0 a_0^4}{a^4} = q_0 \frac{H_0^2 a_0^4}{a^4}.$$

Using this in the equation of motion (6.8.2) to eliminate u_r, we obtain

$$\dot{a}^2 - \frac{q_0 H_0^2 a_0^4}{a^2 c^2} = -1.$$

148

We can eliminate a_0 in favour of q_0 using (6.8.6) evaluated at t_0 to obtain

$$\dot{a} = \left(\frac{q_0 c^2}{(q_0-1)^2 H_0^2} - a^2\right)^{1/2} \frac{1}{a}. \tag{6.8.9}$$

Consequently, following the same argument as in the case of the dust model (§6.7),

$$t = \int_0^t \frac{dt}{da}\, da = \int_0^a \frac{a\, da}{\{q_0 c^2/[(q_0-1)^2 H_0^2] - a^2\}^{1/2}}.$$

To evaluate the integral put

$$a = \frac{q_0^{1/2} c}{(q_0-1)\, H_0} \sin\theta, \tag{6.8.10}$$

so that

$$t/c = \frac{q_0^{1/2}}{(q_0-1)\, H_0} \int_0^\theta \sin\theta\, d\theta = \frac{q_0^{1/2}}{(q_0-1)\, H_0} (1 - \cos\theta). \tag{6.8.11}$$

Equations (6.8.9) and (6.8.10) are parametric equations for the evolution of the scale factor $a(t)$ analogous to equations (6.8.7). In this case, however, we can explicitly eliminate θ to obtain

$$a^2 + \left[t - \frac{q_0^{1/2} c}{H_0(q_0-1)}\right]^2 = \frac{q_0 c^2}{(q_0-1)^2 H_0^2},$$

which is a circle through the origin, centre $a = 0$, $t/c = q_0^{1/2}/H_0(q_0-1)$ (figure 6.6). The lifetime of this universe between the initial and final singularities is

$$T = \frac{2 q_0^{1/2}}{(q_0-1)\, H_0},$$

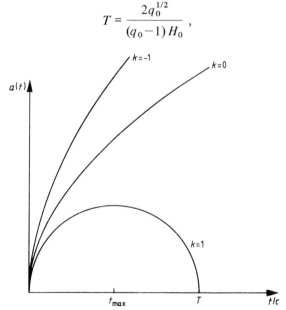

Figure 6.6. Evolution of the scale factor $a(t)$ in the radiation-filled Robertson–Walker models.

149

which is less than that of the corresponding dust model; the radiation universe evolves more rapidly.

(iii) *The case k = −1*. In this case the analogue of (6.8.9) is

$$\dot{a} = \left(a^2 + \frac{q_0 c^2}{(1-q_0)^2 H_0^2}\right)^{1/2},$$

(6.8.12)

and we obtain the parametric representation

$$a = \frac{q_0^{1/2} c}{(1-q_0) H_0} \sinh \theta$$

$$t/c = \frac{q_0^{1/2}}{(1-q_0) H_0} (1 - \cosh \theta).$$

(6.8.13)

Elimination of θ gives the equation of a hyperbola;

$$a^2 - \left[t - \frac{q_0^{1/2} c}{H_0(1-q_0)}\right]^2 = \frac{q_0 c^2}{(q_0-1)^2 H_0^2}.$$

From (6.8.12) we see that the potential energy term dominates the space curvature if

$$a^2 \ll \frac{q_0 c^2}{(1-q_0)^2 H_0^2},$$

or, using (6.8.6) at t_0, if

$$\left(\frac{a_0}{a}\right)^2 \gg \frac{1-q_0}{q_0}.$$

We can translate this into a condition on the temperature in the case of a present 3 K background radiation. Equation (6.8.5) gives

$$q_0 = \frac{8\pi G}{3} \frac{u_r(t_0)}{c^2 H_0^2} \sim 10^{-6}.$$

Then

$$\frac{T}{T_0} = \frac{a_0}{a} \gg 10^3$$

is the required condition. The curvature of space could be important for the dynamics only at temperatures less than 3×10^3 K; but at this temperature the universe is already matter-dominated, and so the radiation model is anyway inapplicable.

6.9. Other Models

The real Universe is composed neither of dust nor of radiation but, at the very least, of a mixture of the two. The dynamics of a non-interacting mixture of

150

radiation and matter can be found explicitly and, not surprisingly, is approximated well in the early stages by a pure radiation model, and at later stages by a dust model, the relativity short, smooth transition occurring around the time t_{eq}. However, the model of a non-interacting brew of radiation and matter is appropriate only in the later stages of evolution, where 'point' galaxies are not much disturbed by the dilute radiation and fill so little of space that they are not able to disturb the radiation either. In fact, in Chapter 8 we shall find that despite the interaction between radiation and matter at early times, the pure radiation model is then a good approximation. The transition between the two approximate regimes presents interesting and difficult physical problems, but is of little consequence for the dynamics of expansion since it represents a negligibly short phase.

It is possible to take account of the random motion of matter, for example clusters of galaxies, by treating a universe filled with a perfect gas at finite temperature expanding adiabatically. The adiabatic condition is equivalent to

$$p \propto \rho^\gamma, \tag{6.9.1}$$

and for a perfect gas we have

$$T \propto p/\rho.$$

In order to apply the field equations we need an equation of state giving μ as a function of p. For a gas of Newtonian particles, of course, $\mu = \rho \propto p^{1/\gamma}$. In the case of a relativistic gas, often considered by cosmologists, we obtain the required relation from (6.9.1) and the conservation equations. Since $\rho \propto a^{-3}(t)$ by conservation of mass, the conservation equation (6.6.7) can be written in terms of ρ instead of $a(t)$, as

$$\rho \, \frac{d\mu}{d\rho} - \mu = p/c^2.$$

Equation (6.9.1) then yields

$$p \, \frac{d\mu}{dp} - \mu/\gamma = p/\gamma c^2$$

as a differential equation for $\mu(p)$. The 'boundary condition' is $\mu \to \rho$ when $p \to 0$, and so we obtain

$$\mu = \rho + \frac{p}{(\gamma - 1) c^2}$$

with ρ given by (6.9.1).

In the literature, discussion is usually restricted either to dust, $p = 0$, or to the highly relativistic case, $\rho \ll p/c^2$, for which $p = (\gamma - 1)\mu c^2$. The requirements that the speed of sound v_s, which can be shown to equal $(dp/d\mu)^{1/2}$, be less than the speed of light and greater than zero impose the restrictions

$$2 > \gamma > 1.$$

The extreme case $v_s = c$ is referred to as 'stiff matter', the possible existence of which in the early universe is purely speculative. For a photon gas we have $\gamma = \frac{4}{3}$ so $v_s = c/\sqrt{3}$.

More exotic possibilities for the equation of state in the early universe have also been considered. There are essentially two extreme versions of what might happen as one goes towards $t = 0$, depending upon whether the number of elementary particles is finite or infinite. If the number is finite, then at sufficiently high temperatures statistical equilibrium is achieved with the creation and destruction of particle–antiparticle pairs proceeding at the same rate. As the temperature is raised further particles become more energetic and their rest mass energy becomes negligible. The system approximates a photon gas with $p = \frac{1}{3}\mu c^2$. The alternative extreme is the creation of more and more massive particles of different types as the energy density increases towards $t = 0$. In this scheme, there are no really elementary particles, but each type of particle is made out of all the others. This is called the 'bootstrap' theory after a certain Baron Münchhausen who saved himself from drowning by pulling himself up by his own bootstraps. In the theory as developed by Hagedorn, there is an upper limit to the temperature attainable by compressing the Universe of somewhat over 10^{12} K. After that the equation of state, and the dynamical evolution, is determined by increasing numbers of more massive particles with relatively negligible kinetic energy. Thus for $t \lesssim 10^{-5}$ s we obtain a system of interacting non-relativistic particles, one possibility for the equation of state of which turns out to be of the form

$$p \propto \frac{\mu}{\ln (\mu/\mu_*)}, \qquad (6.9.2)$$

where $\mu_* \sim 10^{18}$ kg m^{-3}. This will not noticeably alter the dynamics of the late stages of evolution if there is a smooth transition to a radiation-dominated model at 10^{-5} s. This particular theory is probably now of only historical interest and current research on the epoch preceding 10^{-5} s is centred on the Grand Unified Theories (§12.2). Nevertheless, interesting consequences for the growth of perturbations in the early Universe (Chapter 12) could follow from the theory, and this illustrates how cosmology might be used to test our understanding of elementary particle physics.

7. *Cosmological Tests*

7.1. The Classical Tests

The general theory of relativity provides us with a means of predicting the behaviour of any physical system in a gravitational field, provided only that quantum effects can be neglected. In Chapter 5 we described this by saying that general relativity was just special relativity in local, freely falling frames. In particular, we can compute the propagation of electromagnetic radiation through a gravitational field without any *ad hoc* assumptions. General relativity also provides us with a means of computing the gravitational field of a given matter distribution. In particular, we saw in Chapter 6 how it could be used to yield the gravitational field of the smoothed out matter distribution in some approximate models of the Universe.

We can now put these two aspects together and compute the transmission of light through the model universe. This provides us with a means of testing the cosmological theory: we have to find whether any of the reasonable cosmological models fit the picture provided by the observations. If this can be done successfully then we have a means of determining the cosmological parameters of our Universe from the best fits to the observational data. There are three classical tests, so called on the basis of their comparative venerability, which we examine in this chapter. These involve the apparent angular diameter, the apparent magnitude and the observed number density of distant objects. In addition we shall note again the luminosity–volume test first mentioned in Chapter 1.

Departures from the straightforward Newtonian theory enter in two ways: first through the special relativistic effects of the redshift (the diminution in energy and arrival rate of photons), for which Newtonian theory would provide only a first approximation; and secondly, through the curved geometry of space–time, which produces a deviation from exact linearity in the Hubble law. At this level, the Universe differs from the Newtonian expanding box, which provides only a first approximation.

Given a complete disregard for physics, or, equivalently, dropping the assumption that we know anything about physical laws elsewhere in the Universe, we can account for the outcome of these classical tests by means of many *ad hoc* cosmological theories. Usually this involves one extra arbitrary assumption per test. In this way nothing of interest about the world can be derived from cosmology. Granted, however, a modicum of physical knowledge, many competing 'theories' of the Universe, which correctly account for the Hubble law, can nevertheless be ruled out through their failure to pass the classical tests. This underlines the importance of what we have called the fundamental speculation of theoretical cosmology (§5.1).

7.2. The Angular Diameter-Redshift Test

Let an observer be situated at the origin of the coordinate system in a Robertson–Walker universe with metric (6.4.6). For convenience let him use polar coordinates in space (r, θ, ϕ), rather than Cartesian coordinates, so that the metric takes the form

$$d\tau^2 = dt^2 - \frac{a^2(t)}{\psi^2(r)} \, (dr^2 + r^2 \, d\theta^2 + r^2 \sin^2\theta \, d\phi^2).$$

where $\psi(r) = (1 + kr^2/4)$. Let our observer measure the small apparent angular size, $\Delta\theta$, of an object in the direction θ, a galaxy along its major axis, say, or the separation of the lobes of a double radio source. In the absence of gravity, such a galaxy at a distance d would have a linear extent

$$\ell = d \, \Delta\theta,$$

for sufficiently small $\Delta\theta$. We want to obtain a corresponding relation in a Robertson–Walker universe. However, since we cannot attribute a direct operational meaning to the distance in a curved geometry, we look instead for a relation involving the directly measurable reshifts.

   From the geodesic equations it can be shown that θ = constant along an initially radial null geodesic. In fact this follows also by symmetry, since otherwise there would be a preferred direction in space. Thus the difference in θ-cordinates of the ends of the galaxy is $\Delta\theta$. This corresponds to a locally measured linear extent, or proper size, for the galaxy of

$$\ell = \frac{a(t_e) \, r_e \, \Delta\theta}{\psi(r)},$$

where t_e is the time of emission of light from the galaxy at radial coordinate r_e (figure 7.1). This is the separation of its ends in a local freely falling coordinate system at (t_e, r_e, θ, ϕ); therefore

$$\Delta\theta = \frac{\ell\psi(r_e)}{a(t_e) \, r_e} = \frac{\ell(1 + z)}{a_0 [r_e/\psi(r_e)]}. \tag{7.2.1}$$

Consequently, we must compute $r/\psi(r)$, at the galaxy, as a function of its redshift, z.

   Along an incoming null ray we have

$$dt = -\frac{a \, dr}{\psi}. \tag{7.2.2}$$

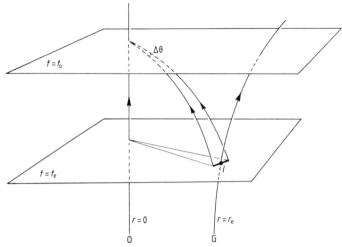

Figure 7.1. Space-time diagram showing the rays from the edges of a galaxy G, emitted at time t_e, observed by an observer O at time t_0.

This follows since $d\lambda = -\,d\xi$ in geodesic coordinates along a radial light ray, or from the condition that $d\tau^2 = 0$ for a null trajectory. Now, from §6.5, we have

$$\frac{dr}{\psi} = \frac{dR}{(1 - kR^2)^{1/2}} = d\chi,$$

and hence

$$\frac{r}{\psi} = R = \frac{\sin k^{1/2}\chi}{k^{1/2}}. \tag{7.2.3}$$

This holds for all the cases $k = 0, \pm 1$, provided we recall that

$$\lim_{k \to 0} = \frac{\sin k^{1/2}\chi}{k^{1/2}} = \chi$$

and $-i \sin i\chi = \sinh \chi$. Equation (7.2.2) can now be written as

$$\int_{t_e}^{t_0} \frac{dt}{a(t)} = -\int_{\chi_e}^{0} d\chi = \chi_e, \tag{7.2.4}$$

where $\chi = \chi_e$ at the emitting galaxy. Thus to find r/ψ we evaluate the integral in (7.2.4) as a function of z, and substitute in (7.2.3). For example, from equation (6.7.19), which gives us the relation between t and z in the $p = 0$ models, we have

$$\int \frac{dt}{a} = \frac{1}{a_0} \int \frac{dt}{dz} \cdot (1 + z)\, dz = -\frac{c}{H_0 a_0} \int (1 + z)^{-1}(1 + 2q_0 z)^{-1/2}\, dz.$$

155

The integration is readily carried out using the substitution $y = (1 + z)^{-1}$; using equation (6.7.5) evaluated at t_0 for $k^{1/2}$, and equation (6.7.6) yields

$$\chi_e = k^{-1/2} \sin^{-1} \left\{ \left(\frac{k^{1/2}c}{a_0} \right) \frac{q_0 z + (q_0 - 1)\,[(2q_0 z + 1)^{1/2} - 1]}{q_0^2 z(1 + z)\,H_0} \right\}. \tag{7.2.5}$$

We obtain the angular diameter-redshift relation in these dust models from (7.2.1):

$$\Delta\theta = \frac{\ell}{(cz/H_0)}\,\{zq_0^2(1 + z)^2\,[q_0 z + (q_0 - 1)((1 + 2q_0 z)^{1/2} - 1)]^{-1}\}.$$

The expression in curly brackets represents the general relativistic correction to the Newtonian approximation. To order z in this correction factor, which is the limit of current observational possibility, we have

$$\Delta\theta = \frac{H_0}{cz}\,\{1 + \tfrac{1}{2}(q_0 + 3)\,z + 0(z^2)\} \tag{7.2.6}$$

The deviation from linearity of $(\Delta\theta)^{-1}$ with z for objects of the same intrinsic diameter yields q_0. Some results are shown in figure 7.2.

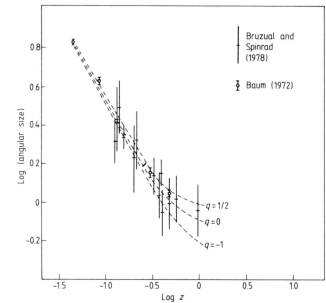

Figure 7.2. The angular diameter-redshift test. Data for galaxies (Baum 1972) and the data for clusters of galaxies with $z > 0.1$ (from Bruzual and Spinrad 1978) are plotted. Closed models would fall above the curve marked $q = \tfrac{1}{2}$, and open ones between $q = 0$ and $q = \tfrac{1}{2}$. The curve $q = -1$ is the steady state model. Best fits to the data are $q_0 = 0.3$ (± 0.2) (Baum) and $q_0 = 0.25$ (± 0.5) (Bruzual and Spinrad).

156

Observations of galaxies are complicated by the fact that they do not have sharp edges. The diameter of a galaxy out to a given relative surface brightness (its isophotal diameter) depends on its redshift, and this makes the test more difficult to apply. The agreement between the results of Baum and those of Bruzual and Spinrad in figure 7.2 may therefore be fortuitous.

Note that it follows from (7.2.6) that θ reaches a minimum for some finite z, after which it increases monotonically. This results from the influence of space-time curvature on the deviation of null geodesics, and is easily seen in the analogy of figure 7.3. For $q_0 = \frac{1}{2}$ (the Einstein–de Sitter model), the minimum angular diameter occurs at a redshift of $5/4$.

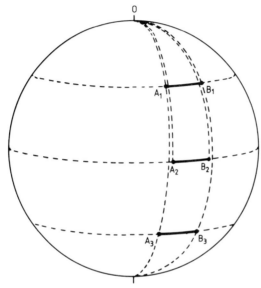

Figure 7.3. The 'rods' A_1B_1, A_2B_2 and A_3B_3 are the same linear size. The angle subtended at the 'observer' (O) decreases as A_1B_1 is moved to the equator (A_2B_2), but then increases again so that A_3B_3 subtends the same angle as A_1B_1.

Once $\Delta\theta$ becomes large, as $z \to \infty$, the formula (7.2.6) is no longer appropriate, since it was derived on the basis of a small $\Delta\theta$. However, a small $\Delta\theta$ is required only so that we do not need to distinguish between an arc, $r = $ constant, and a chord in equation (7.2.1). We can see qualitatively that the angular size of an object must increase as we imagine its redshift to increase; $\Delta\theta > 2\pi$ at some z means that an object of the given size ℓ will not fit inside the visible Universe at this epoch. As we look to higher redshifts, therefore, smaller and smaller objects fill the whole sky. This is because at these times the objects which emitted light we now see were crowded in upon us. As $z \to \infty$ we must have $\ell \to 0$ for finite $\Delta\theta$, and the mathematical point which represents the Universe at its creation is seen by any later observer to fill the whole sky. Thus, at least in principle, whichever way we look out at the sky, we see the big bang. This, of course, is in complete contrast to the

Newtonian picture of objects hurled out into space from some central point fixed in space. In this sense, in general relativity it is space itself that expands.

7.3. The Apparent Magnitude–Redshift Test

A magnitude–redshift relation in Euclidean space was derived in §2.3. The curvature of space–time modifies this result in two ways. First, the intensity of the radiation changes along a ray as a consequence of the redshift effect. Secondly, the rays are bent so that light emitted from a point source is concentrated into an ever-changing bundle of directions. To arrive at an observed flux we have to integrate the received intensity of the radiation over the solid angle subtended by the rays at a unit area located at the point of observation. For this we shall need the relation between the received intensity and that at emission, and an expression for the solid angle into which the received radiation was emitted.

Any point along a light ray can be labelled as usual by the redshift, z, at which a source at that point would appear to some fixed observer. In §8.1 we shall show that the intensity of radiation at frequency v, i_v, varies along the ray in such a way that $i_v/v^3 =$ constant. Since the frequency varies as $v = v_0(1 + z)$, where v_0 is the fixed frequency selected for observation, we have

$$i_{v_0}(0) = (1 + z)^{-3} i_v(z) = (1 + z)^{-3} i_{v_0(1+z)}(z) \qquad (7.3.1)$$

as the relation between the observed intensity, $i_{v_0}(0)$, and that emitted at z at the appropriate frequency, $i_{v_0(1+z)}(z)$. This result is the analogue of the constancy of i_v along a ray in the absence of gravity.

Consider now the observed flux. To obtain this we place a unit area normal to the beam and integrate the intensity over the solid angle subtended by rays crossing this area (figure 7.4). Therefore

$$f_{v_0} = \int i_{v_0} \cos \theta \; d\Omega \approx i_{v_0} \; d\Omega_0 \qquad (7.3.2)$$

for a source subtending a sufficiently small solid angle. In order to relate this to the energy output of the source, we need to know the solid angle at the source, $d\Omega_e$, into which the received radiation flows. The deviation of geodesics means that the angles $d\Omega_e$ and $d\Omega_0$ are not equal. The relation between $d\Omega_0$ and $d\Omega_e$ is called the reciprocity theorem, which we now derive for the special case of a Robertson–Walker cosmology (see Ellis 1971).

Consider rays emitted from a source at the coordinate origin in the range $d\theta$, $d\phi$ about some direction (θ, ϕ), and received at (t_0, r). The 'proper' area of the bundle of rays, as measured locally at reception by a freely falling observer, is

$$dS_0 = \frac{a_0^2}{\psi^2(r)} r^2 \; d\Omega_e. \qquad (7.3.3)$$

Now follow the cone of rays $d\Omega_0$ (figure 7.4) back in time to the point of emission.

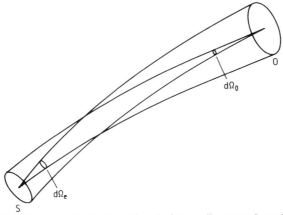

Figure 7.4. Null rays emitted into solid angle $d\Omega_e$ at the source S, are incident on unit area of surface at the observer O. But the solid angle that can be measured is that subtended at the observer by the source, $d\Omega_0$.

In this time-reversed model we must have

$$dS_e = \frac{a_e^2}{\psi^2(r)} r^2 \, d\Omega_0. \tag{7.3.4}$$

It follows that for unit areas at the source and observer we must have

$$d\Omega_0 = d\Omega_e (1 + z)^2 \tag{7.3.5}$$

which is the reciprocity theorem. One can understand this as the special relativistic dilation of solid angle by relative motion holding here, however, over arbitrary separations of source and observer.

Putting the results from equations (7.3.1), (7.3.3), and (7.3.5) together gives

$$f_{\nu_0} = \frac{i_{\nu_e} \, d\Omega_e}{(1+z)} = \frac{\ell_{\nu_e}}{4\pi} \cdot \frac{1}{(1+z)} \cdot \left(\frac{\psi(r)}{a_0 r}\right)^2 \tag{7.3.6}$$

if the source emits isotropically, where ℓ_{ν_e} is the total power per unit band width (W Hz^{-1}) emitted at frequency $\nu_e = \nu_0(1+z)$. For the total (bolometric) flux we have

$$F = \int f_{\nu_0} \, d\nu_0 = \frac{1}{4\pi} \frac{1}{(1+z)^2} \left(\frac{\psi(r)}{a_0 r}\right)^2 \int \ell_{\nu_e} \, d\nu_e. \tag{7.3.7}$$

We can think of this as the energy emitted into the appropriate solid angle at the source reduced by one factor of $(1 + z)$ to allow for the effect of the redshift on the energy of the photons, and a second factor to allow for the effect of time-dilation on the rate at which photons are received.

Multiplying equation (7.3.6) by f_{ν_e}/f_{ν_0}, and writing ν for ν_e, gives

$$f_\nu = \frac{\ell_\nu}{4\pi} \cdot \frac{1}{(1+z)^2} \cdot \left(\frac{\psi(r)}{a_0 r}\right)^2 \{(1+z) f_\nu/f_{\nu/1+z}\}. \tag{7.3.8}$$

159

Apart from the factor in curly brackets, equation (7.3.8) has the same form as equation (7.3.7).

It is usual to express the result in terms of (astronomical) magnitudes. The absolute magnitude at frequency ν is, by definition, the magnitude at a distance of 10 pc,

$$M_\nu = -2.5 \log_{10} \left(\frac{\ell_\nu}{4\pi (10 \text{ pc})^2} \right).$$

The apparent magnitude is

$$m_\nu = -2.5 \log_{10} f_\nu.$$

Consequently, using equation (7.3.8),

$$m_\nu = M_\nu + 5 \log_{10} \left\{ (1 + z) \frac{a_0 r}{\psi(r)} \cdot \frac{1}{(10 \text{ pc})} \right\} + 2.5 \log_{10} \left(\frac{f_{\nu/1+z}}{(1 + z) f_\nu} \right).$$

The final term is the K-correction: it expresses the difference between this result using magnitudes based on small band widths, and the analogous result derived from (7.3.7) for bolometric magnitudes. Radio sources typically have power law spectra, $f \propto \nu^{-\alpha}$, so the K-correction for these is $2.5(\alpha - 1)$.

As in §7.2, for a specific cosmological model we can eliminate $a_0 r/\psi = a_0 \sin k^{1/2} \chi_e / k^{1/2}$ (equation 7.2.5) in terms of the standard parameters, which can therefore be determined by the deviation of $m - \log z$ from a straight line.

Neglecting corrections of order z^2 and higher, we obtain

$$m_\nu = M_\nu - 5 + 5 \log cz/H_0 + (15/2)(1 - q_0) z + K_\nu.$$

The data of figure 2.2 give $q_0 = 1.6 \pm 0.4$, but this requires an (uncertain) correction for changes in the luminosities of galaxies with time. From similar data, Gunn and Oke (1975) find $-1.27 \leqslant q_0 \leqslant 0.33$.

7.4. The Number–Luminosity Test

In Chapter 1 we looked at counts of galaxies as functions of limiting apparent magnitude in a Euclidean space model. The number–luminosity test takes into account the effect of curvature on the propagation of light.

An element of volume $dS_e \, d\xi$, measured at P, is observed at O to contain $dN = n \, dS_e \, d\xi$ galaxies, where n is the locally measured number of galaxies per unit volume at P (figure 7.5). The element of area dS_e subtends a solid angle $d\Omega_0$ at O. From equation (7.3.4), we have

$$dS_e = a_e^2 \left(\frac{r}{\psi(r)} \right)^2 d\Omega_0.$$

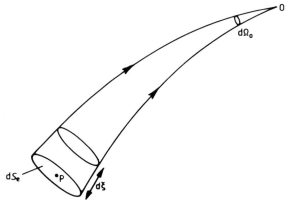

Figure 7.5. The element area dS_e at P subtends a solid angle $d\Omega_0$ at the observer O. The diagram represents the paths of null rays projected into the three-dimensional space of the observer, $t = t_0$.

As usual, the proper length $d\xi$ is $a_e\, dr/\psi(r)$ in the coordinates based on O, so

$$dN = n\, \frac{a_e^2 r^2}{\psi^2}\, d\Omega_0\, \frac{a_e\, dr}{\psi}.$$

Then

$$N(\chi_e) = 4\pi \int_0^{\chi_e} n a_e^3 R^2(\chi)\, d\chi$$

is the total number of sources out to χ_e.

Assume now that the sources have the same intrinsic power and that their number is conserved. Then $n a_e^3 = n_0 a_0^3$, where n_0 is the source density at the observer. Both of these assumptions can be readily dropped. For variable source power we simply apply the theory to each source type and sum over the different types. If n is not conserved it is only necessary to specify how it changes as a function of z, and to include this in the integration. For our simple case we obtain

$$N(\chi_e) = 4\pi n_0 a_0^3 \int_0^{\chi_e} \frac{\sin^2 k^{1/2}\chi}{k}\, d\chi$$

$$= \begin{cases} \dfrac{4\pi n_0 a_0^3}{2k}\left[\chi - \dfrac{\sin 2k^{1/2}\chi}{2k^{1/2}}\right] & k \neq 0 \\[4mm] \dfrac{4\pi n_0 a_0^3}{3}\, \chi_e^3. & k = 0. \end{cases} \qquad (7.4.1)$$

The value of χ_e is determined by the limiting magnitude to which the observations are carried. Let this be given by a specific flux $f_\nu = S$. For radio sources of

161

intrinsic power ℓ_ν per unit frequency, having spectral flux $f_\nu \propto \nu^{-\alpha}$, equation (7.3.8) gives

$$\ell_\nu = 4\pi a_0^2 \chi_e^2 (1+z)^{1+\alpha} S \qquad (7.4.2)$$

as the relation between the unknown χ_e and the observed flux S. The redshift z is given as a function of χ_e by equation (7.2.5), but unfortunately we cannot eliminate χ_e directly between equations (7.4.2) and (7.4.1). Nevertheless, we can see the qualitative form of the departure from the Euclidean result $\ell_\nu \propto S^{-3/2}$.

Take, for simplicity, the $k = 0$ case. Then

$$N = \frac{4\pi n_0}{3} \left(\frac{\ell_\nu}{4\pi S} \right)^{3/2} (1+z)^{-3(1+\alpha)/2}$$

from equations (7.4.1) and (7.4.2). As S decreases, so z must increase, and the effect of the extra factor $(1+z)^{-3(1+\alpha)/2}$ is to decrease N. The slope of the $\log N$-$\log S$ relation is therefore flatter than the Euclidean slope of $-3/2$ at fainter specific fluxes. It can be shown that this result is valid for all reasonable cosmological models (assuming $\Lambda = 0$).

We can understand what is happening here in terms of our simple picture of the expanding Universe. There are three effects: (i) The light from distant galaxies is redshifted, so the sources appear fainter than they would otherwise. This moves a point on the $\log N$-$\log S$ curve (figure 7.6) to the left, so flattening the slope. (ii) The sources were closer together at the time in the past when their radiation was emitted. We are therefore seeing now a picture of a more crowded universe. This raises a point at given S on the $\log N$-$\log S$ curve, thereby increasing the slope. (iii) We observe in a fixed band width radiation that was emitted at increasingly higher frequencies as we look to fainter sources, whereas the Euclidean model assumes that the observed and emitted band width are the same. The brightness of radio sources decreases in general with increasing frequency, and this leads to a small amount of flattening of the $\log N$-$\log S$ curve for fainter sources. From our detailed discussion of the $k = 0$ case, we see that effect (i) must dominate.

This test is usually carried out for radio sources because radio surveys can go to greater depths than optical counts, and the counting problem is much simplified, since essentially all radio sources at high Galactic latitude are extragalactic. For reference to the literature note that radio astronomers use the term 'flux density' for the specific flux S.

The history of this test is well known. The original Cambridge radio surveys showed an initial steepening of the curve towards lower flux density. This rules out the steady state theory which predicts a flattening, and at first glance, it also rules out the general relativistic models. However, in the evolving models one is allowed to drop the assumptions of conservation of sources and their absolute luminosities. We can allow either an increased number of radio sources in the past (density evolution) or an increased power for a given number of radio sources (luminosity evolution), or any combination of the two, sufficient to fit the data and physically

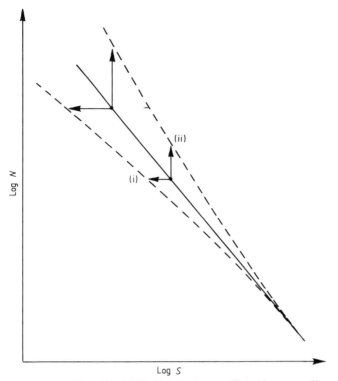

Figure 7.6. The effect of redshift (i) and the crowding of sources (ii), on the log N–log S curve.

quite reasonable. This avenue is not open to the steady state theorists, since this model must have the same properties at all times. However, the interpretation of the early surveys was open to dispute, and was indeed much disputed. They turned out to be 'confusion limited' since close sources were counted as single. Nevertheless, more recent surveys (figure 7.7) also show the initial steepening towards low flux density, and consequently add weight to the evidence against the steady state theory.

A more recent test related to the log N–log S counts is the luminosity–volume test (Chapter 1), which can be used when redshifts are available for a complete sample of objects above a limiting flux density. Let V be the volume of space within the redshift z of a source, and let V' be the volume within which the source could have been and be still observed. Then

$$\frac{V}{V'} = \frac{4\pi \int_0^{\chi_e} a_0^3 R^2 \, d\chi}{4\pi \int_0^{\chi_e'} a_0^3 R^2 \, d\chi}$$

where χ_e is determined by equation (7.2.5) and χ_e' by

$$\ell_\nu = 4\pi R^2(\chi_e') \, [1 + z(\chi_e')]^{1+\alpha} \, S.$$

163

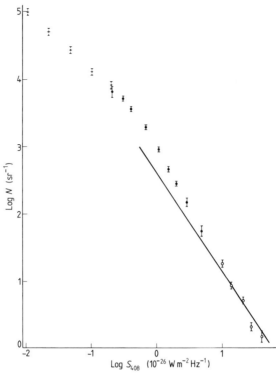

Figure 7.7. The log N–log S relation for radio sources at 408 MHz. The data are taken from Robertson (1973) (open circles), the B2 Survey (Colla *et al* 1972) (full circles), and the 5C Survey (Pooley and Ryle 1968) (crosses). The flux is in units of 10^{-26} W m^{-2} Hz^{-1} = 1 Jansky (adapted from Grueff and Vigotti 1977). The solid line has a slope -1.5. The data show clear evidence for a steeper slope as one moves to lower flux densities (S), flattening again towards the faintest sources.

The mean value, $\langle V/V' \rangle$, of V/V' for a uniform distribution of sources turns out to be 0.5. If $\langle V/V' \rangle$ is greater than 0.5 for the sample this implies that a majority of the sources lie at large redshifts. Applied to samples of quasars, this test provides stong evidence of evolution. It is a topic of current research to determine the extent of this evolution in the form of an expression for the density of quasars $\rho(z)$ at a redshift z. Popular fits to the data are power law evolution, $\rho(z) \propto (1 + z)^{\beta}$ with $\beta \sim 6$, or exponential evolution $\rho(t) \propto \exp(-t/t_0)$. In this way this test becomes not so much a trial for cosmology, but a means of extracting restrictions on compatible astrophysics.

7.5. Alternative Cosmologies

The classical tests have not been at all successful in providing us with the parameters of a satisfactory general relativistic model, since a large range of possibilities are compatible with the available data. One might therefore permit oneself to

164

wonder whether the tests test anything. They do, of course, since, for example, extreme values of q_0 are not possible, and the steady state theory has been eliminated. But more than this, the tests are very good at ruling out *ad hoc* cosmological models, provided only that one admits a modest knowledge of physics as a constraint on such speculative edifices. Of course, no finite number of tests can rule out 'theories' which admit an arbitrary number of adjustable parameters. We shall illustrate this point with two examples.

The first of these is a 'tired light' model. One imagines a static Euclidean universe in which light loses energy in its journey to us from the distant galaxies because of, say, some small correction to Maxwell's equations. It is not too difficult to account for the Hubble law locally in this theory. We can postulate

$$-\frac{dE}{E} = -\frac{d\nu}{\nu} = \frac{H_0}{c}\,dr$$

as the relation between frequency change $d\nu$ and distance dr. On integration, we get

$$r = c/H_0 \log(\nu_e/\nu_0) = c/H_0 \log(1+z) = cz/H_0 + \ldots$$

and Hubble's law is obtained to first order.

It is now easy to show that this conflicts with the angular diameter test. For

$$\Delta\theta = \frac{\ell}{r} \approx \frac{\ell H_0}{c}\frac{1}{z}(1 - z/2).$$

This is equivalent to a general relativistic dust model with $q_0 = -4$, well outside the large range of reasonable possibilities.

As a second example consider a static Euclidean universe in which the masses of elementary particles increases with time. Again it is easy to satisfy the Hubble law. We put

$$m(t) = \frac{ma(t)}{a(t_0)};$$

then, according to the Bohr model of the atom, the frequency of a spectral line is proportional to the electron mass, so

$$1 + z = \frac{m(t_0)}{m(t_e)} = \frac{a(t_0)}{a(t_e)}. \tag{7.5.1}$$

This yields Hubble's law as a first approximation in the usual way. To a second approximation we have

$$\frac{1}{1+z} = 1 - \frac{H_0 r}{c} + \frac{1}{2}q_0\left(\frac{Hr}{c}\right)^2 + O(r^3 H^3/c^3) \tag{7.5.2}$$

by Taylor expansion of (7.5.1). Here r is the Euclidean distance of a source, $r = c(t_0 - t_e)$.

To satisfy the classical tests we might postulate a concomitant change in the luminosities and sizes of sources. Suppose the absolute luminosity, L, of a galaxy is related to its mass, M, by $L \propto M^\beta$, for some constant β, and let us look at the $\log N$–$\log S$ relation. We have

$$S = \frac{L_0(1+z)^{-\beta}}{4\pi r^2} \frac{1}{(1+z)}, \qquad N = \tfrac{4}{3}\pi r^3 n,$$

where the extra factor of $(1+z)^{-1}$ takes account of the effect of the redshift on the observed photon energy, and n is the constant source density. We eliminate z using (7.5.2) to get

$$\log N = \text{constant} - \tfrac{3}{2} \log S - \frac{3(1+\beta) H_0}{c} \left(\frac{L_0}{4\pi}\right)^{1/2} S^{-1/2} + O(S^{-1}).$$

This yields a log N–log S curve which steepens for smaller S, in agreement with observations, if $(1+\beta) < 0$. So if sources were brighter in the past, the observed steepening could be accounted for.

However, this is the the wrong conclusion because it disregards our knowledge of stellar structure. From observation and theory we know that the luminosity of stars decreases with decreasing mass, so that, in the proposed cosmology, galaxies must have been fainter in the past, not brighter. The observations therefore rule out the model in this form.

The importance of this discussion is not simply that two particular models can be ruled out. The moral to be drawn is that general relativity is not loved for a beauty and completeness alone, that blinds cosmologists to alternative theories. Alternatives can be and have been tried and have been found wanting.

8. Matter and Radiation

8.1. Radiation in an Expanding Universe

The behaviour of electromagnetic radiation in an expanding universe can be obtained from Maxwell's equations if we assume, in accordance with the Principle of Equivalence, that these equations are valid locally in freely falling frames. Alternatively, we can adopt a kinetic theory approach and describe the radiation by means of a distribution function which specifies the number of photons in each mode at each point. It can be shown that the two approaches give exactly the same results in cases where the wave nature of light is not important. Here we adopt the second alternative as the simpler.

Consider a swarm of photons with energies between E and $E + \Delta E$ moving through a space–time with the Robertson–Walker metric (6.4.6). At time t, let them move in a cone of solid angle $\Delta\Omega$ about the direction $\mathbf{n}$, and be spread through a proper volume $\Delta V = \Delta\xi_1 \Delta\xi_2 \Delta\xi_3$ about a freely falling fundamental observer O (figure 8.1). At some small time δt later these photons occupy a volume $\Delta V'$ about a similar observer O$'$, a distance $c\,\delta t$ from O. O$'$ will measure an energy spread $\Delta E' \sim \Delta E a(t)/a(t + \delta t)$ about $Ea(t)/a(t + \delta t)$ as a result of the usual redshift, and he will measure a proper volume

$$\Delta V' \sim [a(t + \delta t)\, \psi(\mathrm{O})/a(t)\, \psi(\mathrm{O}')]^3\, \Delta V$$

because of the expansion of the universe. Since $[\psi(\mathrm{O})/\psi(\mathrm{O}') - 1]$ is of order $(\delta t)^2$ we can set $\psi(\mathrm{O})/\psi(\mathrm{O}') \sim 1$.

The direction of motion $\mathbf{n}'$ measured by O$'$ is given by the geodesic equation, but we do not need to evaluate it explicitly. The solid angle into which photons are moving locally is constant along a ray; this follows either from the symmetry of the model, or, equivalently, because the radial null rays have θ = constant and ϕ = constant, as in special relativity. This should not be confused with the solid angle subtended by a distant source, which was discussed in §7.2. The element of volume in the photon momentum space is

$$\Delta V_{\mathrm{p}} = p^2\, \Delta p\, \Delta\Omega = (1/c^3)\, \Delta E \cdot E^2\, \Delta\Omega \text{ at O,}$$

and

$$\Delta V_{\mathrm{p}}' = [a(t)/a(t + \delta t)]^3\, \Delta V_{\mathrm{p}} \text{ at O}'.$$

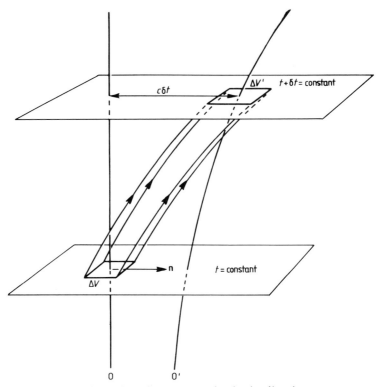

Figure 8.1. Photons from the volume ΔV moving in the direction $\mathbf{n}$ occupy a volume $\Delta V'$ at time $t + \delta t$.

Therefore the product $\Delta V \Delta V_p$ is constant along the ray from O to O'. The expansion of the universe leads to an increase in the volume occupied by a given set of photons, but their spread in momentum is reduced because of the redshift in frequency, and the two effects just balance. This is a special case of Liouville's theorem, which states that the volume of phase space occupied by a given set of freely propagating photons is constant in time.

We now introduce the distribution function to describe a gas of photons. Let $f(t, \mathbf{x}, E, \mathbf{n}) \Delta V \Delta V_p$ be the probability at time t of finding a photon in the volume ΔV about the point $\mathbf{x}$, with momentum in the range ΔV_p about $\mathbf{p} = E/c\,\mathbf{n} = h\nu/c\,\mathbf{n}$. At time $t + \delta t$ such a photon will have moved to $\mathbf{x} + \delta \mathbf{x}$ and will be travelling in the direction $\mathbf{n} + \delta \mathbf{n}$ with energy $E + \delta E$, where the increments are restricted to be compatible with the geodesic equation of the motion of the photon. Now, if we ignore the possibility of the creation and destruction of photons by emission and absorption processes, the probability of finding a photon in this new state at $t + \delta t$ must equal that of finding it in the original state at t. Therefore

$$f \Delta V \Delta V_p |_t = f \Delta V \Delta V_p |_{t + \delta t}.$$

But, according to Liouville's theorem, the volume of phase space is unchanged, so

$\Delta V(t) \Delta V_{\mathrm{p}}(t) = \Delta V(t + \delta t) \Delta V_{\mathrm{p}}(t + \delta t)$. Consequently,

$$f(t, \mathbf{x}, E, \mathbf{n}) = f(t + \delta t, \mathbf{x} + \delta \mathbf{x}, E, \mathbf{n} + \delta \mathbf{n}).$$

This is a form of the kinetic equation for non-interacting photons. We can express the results by saying that f is constant along a ray.

We want to re-express this result in terms of the intensity of the radiation field in place of the distribution function. This will yield the equation of radiative transfer.

Let the total number of photons be N. Statistical mechanics tells us that there is one photon state in a volume h^3 of phase space. The average number of photons in volume dV with energy between E and $E + dE$ travelling into $d\Omega$ is therefore $Nf\,dV E^2\,dE\,d\Omega/h^3 c^3$. Therefore the energy per unit volume flowing into solid angle $d\Omega$ is $U_\nu\,d\nu\,d\Omega = h\nu \cdot Nf\nu^2\,d\nu\,d\Omega/c^3$, since $\nu = E/h$. The intensity i_ν is related to the energy density per unit solid angle per unit frequency, U_ν, by $i_\nu = cU_\nu$, so

$$i_\nu = \frac{h\nu^3}{c^2}\,Nf.$$

Therefore the constancy of f along a ray implies that i_ν/ν^3 is constant.

Suppose that an observer in a Robertson–Walker cosmological model receives a beam of radiation at frequency ν_0. At a redshift of z along the beam the radiation has frequency $\nu = (1 + z)\nu_0$. From the constancy of i_ν/ν^3 it follows that the received intensity is related to that at z by

$$i_{\nu_0} = (1 + z)^{-3} i_\nu. \tag{8.1.1}$$

This is the analogue of the constancy of i_ν along a ray in the absence of gravity (e.g. Chandrasekhar 1960).

Equation (8.1.1) can be written in the alternative form of a radiative transfer equation, which we shall need when we consider the inclusion of emission and absorption processes in §8.8. Differentiating equation (8.8.1) we have

$$0 = \frac{\mathrm{d}}{\mathrm{d}z} \{(1 + z)^{-3} i_{\nu_0(1+z)}(z)\},$$

or

$$\frac{\mathrm{d}i_{\nu_0(1+z)}}{\mathrm{d}z} - \frac{3}{1 + z}\,i_{\nu_0(1+z)} = 0. \tag{8.1.2}$$

Note that $i_\nu(z)$ is a function of z explicitly, through the dependence of the intensity on time, and implicitly through the dependence of ν on z. Both of these must be taken into account in the differentiation in equation (8.1.2).

The result (8.1.1) is of a similar form to the relation (3.4.1) between the intensity measured in different frames of reference at the same point. In fact, that result can also be obtained from the invariance of i_ν/ν^3. Note that the demonstration here shows these results to be an exact consequence of general relativity, and not merely an approximation in the Newtonian expanding box. This conclusion is not at all obvious. In taking account of general relativistic effects, one might expect to draw an analogy with the behaviour of light rays in a medium, where a changing refractive index leads to deviations from straight-line trajectories, yet there is no additional correction for a refractive index effect in the transfer equation in general relativity. The explanation is that in a refracting medium the local velocity of light differs from c, whereas in a gravitational field, *in vacuo*, the locally measured light speed is still c.

As long as we can treat electromagnetic radiation as a classical gas of photons, and as long as we can ignore its interaction with matter, the kinetic equation (8.1.2) controls its propagation through the Universe. From this we obtain a simple proof that radiation which is initially black-body subsequently remains black-body in the Robertson–Walker models.

The intensity of black-body radiation is given by the Planck distribution.

$$i_\nu = \frac{2h\nu^3}{c^2} \left[\exp(h\nu/kT) - 1\right]^{-1}. \tag{8.1.3}$$

The transfer equation (8.1.2) expresses the constancy of i_ν/ν^3 along a ray, so

$$\frac{i_{\nu(1+z)}}{\nu^3 (1+z)^3} = \frac{i_\nu}{\nu^3}.$$

Together with equation (8.1.3), this gives

$$\{\exp[h\nu(1+z)/kT(z)] - 1\}^{-1} = [\exp(h\nu/kT) - 1]^{-1},$$

from which the temperature, $T(z)$, at redshift z is

$$T(z) = (1+z)\, T\,(0) = (a_0/a)\, T\,(0).$$

The form of the distribution is preserved; only its temperature changes. From this we derive the evolution of the total energy density,

$$u = aT^4 \propto (1+z)^4,$$

the spectral number density of photons,

$$n_\nu = \frac{4\pi}{c} \frac{i_\nu}{h\nu} \propto (1+z)^2,$$

and so on. In particular, for the total number of photons per unit volume at redshift z we have

$$n(z) = \int n_{\nu_0(1+z)}\, d\nu_0(1+z) \propto (1+z)^3 \propto (a_0/a)^3.$$

The number density is therefore inversely proportional to the volume, in accordance with the conservation of the total number of photons.

8.2. Matter in an Expanding Universe

The behaviour of matter at the microscopic, but still classical level is again described by means of a distribution function. This gives the probability of finding a point particle with a given momentum at a given point, i.e. in a particular region of phase space. As usual, we can think of our point particle either as the atom of a gas in some uniform model universe, or as a cluster of galaxies in some smooth approximation to the present Universe. In either case, it is sufficient to confine ourselves to particles moving relative to fundamental observers with speeds much less than that of light.

To set up the theory we return, as always, to the fundamental postulate of general relativity, which is the Principle of Equivalence. This tells us that in a local freely falling frame we can ignore gravity to first order.

Consider, therefore, a collection of particles moving through a space–time with metric (6.4.6). Let the particles have a spread in velocity $\Delta\mathbf{v} = (\Delta v_1, \Delta v_2, \Delta v_3)$ about $\mathbf{v}$, and be spread through a volume $\Delta V = \Delta\xi_1 \Delta\xi_2 \Delta\xi_3$ about a freely falling observer O (figure 8.2). At some small time δt later these same particles occupy a volume $\Delta V'$ about O', a distance $\mathbf{v}\delta t$ from O. O' will measure a velocity $\mathbf{v}'$ and a

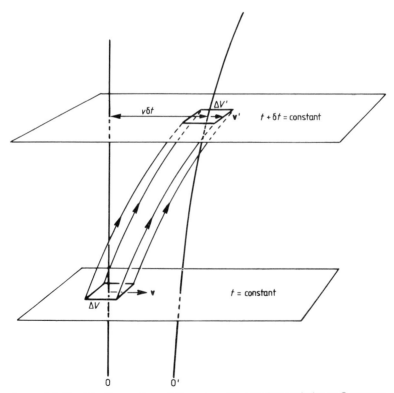

Figure 8.2. Particles from volume ΔV moving with velocity $\mathbf{v}$ relative to O occupy a volume $\Delta V'$ at time $t + \delta t$.

spread $\Delta v'$. We need to relate $\Delta V'$ and $\Delta v'$, as measured by O', to ΔV and Δv. Since O' makes his measurements at $t + dt$ he measures a proper volume $\Delta V' = [a(t + dt) \, \psi(O)/a(t) \, \psi(O')]^3 \, \Delta V$. The ratio $\psi(O)/\psi(O') \sim 1$ to first order in δt, so can be omitted henceforth. Since O' is moving away from O with velocity $\dot{a}(t + dt) \, \xi(t + dt)/a(t + dt)$ (§6.5) he measures a velocity

$$\mathbf{v}(t + dt) = \mathbf{v}(t) - \dot{a}/a \, \xi = \mathbf{v}(t) - \dot{a}(t)/a(t) \, \mathbf{v}(t) \, \delta t,$$

so $\delta v/v = - \delta a/a$ to first order. Therefore $\mathbf{v}(t + \delta t) = [a(t + \delta t)/a(t)]^{-1} \, \mathbf{v}(t)$, and the *spread* in velocities is also reduced by this ratio of the expansion factors.

It follows that the product $\Delta V \Delta v_1 \Delta v_2 \Delta v_3$ is unchanged along the tube of particles. The expansion of the universe increases the volume occupied by a given set of freely moving particles, but their spread in velocity is reduced, since this is measured relative to observers moving in the same direction as the particles; the two effects just balance. This is a special case of Liouville's theorem: the volume of phase space occupied by a given set of free particles does not change with time. It is for this reason that the concept of phase space is important.

If particles are neither created nor destroyed, the probability of finding a particle in an element of phase space mapped out by a swarm of particles does not change as we follow a swarm. This probability is given by the product of the distribution function $f(t, \xi, \mathbf{v})$ and the element of volume of phase space $\Delta V \Delta p_1 \Delta p_2 \Delta p_3 = m^3 \Delta V \Delta v_1 \Delta v_2 \Delta v_3$ for particles of mass m. According to Liouville's theorem, the volume of the element of phase space is constant. Consequently, the distribution function is preserved along a particle path. In symbols,

$$f(t, \xi, \mathbf{v}) = f[t + dt, \xi + \mathbf{v} \, dt, \mathbf{v}(1 - \dot{a}/a) \, dt] \tag{8.2.1}$$

in a local Minkowski coordinate system (t, ξ). Equation (8.2.1) is a version of the kinetic equation for non-interacting particles in an expanding universe. To find out what happens to an initially given distribution of particles we have to solve this equation.

In reality, things are a lot more difficult because we need to include the effect of interactions and, in particular, of collisions between particles. Although we shall not venture to do so here, collisions can be dealt with in a parallel manner to their treatment in the absence of gravity as a consequence of the Equivalence Principle. The important result for us is that in equilibrium the collision term vanishes: there are as many particles nudged into a given phase space element per unit time as are knocked out of it. Of course, this result is required for consistency in the meaning of thermal equilibrium!

Suppose, therefore, that we start at some initial instant with an equilibrium Maxwellian distribution

$$f = \exp[-(\beta/2) \, mv^2 - \mu].$$

We can ask whether the distribution function can remain of this form as the Universe expands, and, if so, how μ and β must change with time. In fact, we shall assume that μ is constant and verify the consistency of this assumption presently. From equation (8.2.1), in the case where f is independent of ξ, and μ is constant, we have

$$\delta f = 0 = \frac{\partial f}{\partial t}\delta t - \frac{\dot{a}v}{a}\delta t\,\frac{\partial f}{\partial v}.$$

So

$$\left(\delta\beta - 2\frac{\delta a}{a}\right)\left(\frac{\beta mv^2}{2}\right) = 0,$$

from which

$$\beta \propto a^2(t). \tag{8.2.2}$$

Now β looks as if it wants to play the role of an inverse temperature, as in the absence of gravity, but we must check that it is exactly this and not different by factors of $a(t)$. Since locally we must recover non-gravitational physics, we must define temperature, as usual, in terms of the average particle energy:

$$u/n = \tfrac{3}{2}kT, \tag{8.2.3}$$

where u is the mean thermal energy density, and n is the mean particle number density. We expect the volume element in velocity space to be $4\pi v^2\,dv$, so

$$n = 4\pi \int_0^\infty fv^2\,dv,$$

and hence

$$\exp(-\mu) = n\left(\frac{m\beta}{2\pi}\right)^{3/2}.$$

From this it follows that the assumption of constant μ is consistent with equation (8.2.2), since we know $n \propto a^{-3}$. This confirms also our choice of volume element. Furthermore, we now have

$$u = 4\pi \int (\tfrac{1}{2}mv^2)\,fv^2\,dv = \tfrac{3}{2}n/\beta. \tag{8.2.4}$$

Comparison of equations (8.2.3) and (8.2.4) yields $1/\beta = kT$. Therefore, from equation (8.2.2), we obtain the final result

$$T \propto a^{-2}(t),$$

or

$$T \propto (1+z)^2,$$

for the dependence of the matter temperature on redshift in a Robertson–Walker universe containing independent point particles.

The result can be readily understood: the velocity of a particle with respect to the local fundamental observer decreases with time, and, since the temperature of a gas is a measure of the mean square velocity, it follows that the gas cools. In fact, we can conclude that the gas behaves as an ideal gas cooling adiabatically, with pressure $p \propto T/V \propto V^{-5/3}$.

From equation (8.2.4) we deduce that the dependence of the thermal energy density on redshift is $u = u_0(1 + z)^5$. For sufficiently large z we see that the thermal energy exceeds the rest mass energy, at which point special relativistic corrections cannot be ignored. The mass of the particles then depends on their velocity, and the distribution function is no longer a homogeneous function of T and v, of the form $f(T^x v^y)$. The shape of the distribution is no longer preserved under expansion, and the population of the states changes. Physically this manifests itself as the appearance of a so-called bulk viscosity, which results in the transfer of mass–energy to random motion as the gas expands, and a consequent increase in entropy. This is a purely special relativistic effect, and can for the most part be neglected. For zero rest mass particles the bulk viscosity is zero, and the effect does not occur. For massive particles it would dominate when $u_0(1+z)^5 \sim n m_0 c^2 \sim n_0 m_0 c^2 (1+z)^3$, which gives, very approximately, $z \gtrsim 10^{10}$. This corresponds to a radiation temperature in excess of 10^{10} K at which the creation of electron–positron pairs represents a much more significant divergence from the assumptions of this section (see §8.4).

8.3. The Interaction of Matter and Radiation

At this point we can begin to relate the abstract models of pure radiation and purely material universes to the interacting reality. At the present epoch the situation is simple because the coupling between matter and radiation is exceedingly weak. A medium of neutral hydrogen would be virtually transparent to black-body radiation peaking in the microwave region. Even a dense intergalactic medium ionised by heating at some stage would allow a photon a 99% chance of making it across the present Universe without being scattered by electrons. Clearly the clusters of galaxies are scarcely perturbed in their modest random motions by the presence of the background radiation. Consequently, in the present Universe the background radiation and the matter each behave as if the other were not there: the matter has its own temperature, which decreases as $T_m \propto a^{-2}$, and the radiation has its temperature which follows $T_r \propto a^{-1}$.

As we look back into the past there comes a stage at which the radiation and matter temperatures are sufficiently high ($\gtrsim 10^4$ K) to ensure almost complete ionisation. The 3 K background would exceed 10^4 K at a redshift of about 3000. This is well before the earliest time at which galaxies could possibly have formed, since there is no room for all the galaxies separately in a universe this young. Therefore, before this time, t_{rec}, we might expect to find the universe filled with a homogeneous plasma. The redshift at which this occurs is estimated more accurately in §8.6. Fully ionised matter of sufficient density and hot radiation are very strongly coupled together. At sufficiently early times, before t_i, say, we find that the radiation and matter can keep in equilibrium, so $T_m = T_r$. After t_i, and until the ions and electrons are completely recombined, the temperature falls at a rate between $(1 + z)^{-2}$ and $(1 + z)^{-1}$ as we go forward in time. In fact, we shall show that the temperature is scarcely perturbed at all by the presence of matter, and that up to t_{rec} we have $T_m \approx T_r$. This occurs despite the fact that for part of this era, the matter rest mass–energy density dominates the energy density of radiation.

This result is obtained by considering the specific heats of the gas and radiation. To raise the temperature of unit mass of matter by ΔT_m requires an input of heat $C_m \Delta T_m$, where C_m is the appropriate specific heat per unit mass. For an order of magnitude estimate we can take $C_m = C_V \sim k/m_p$. The total amount of internal energy of the radiation that could be available to heat unit mass of matter is of order $a T_r^4/n m_p$. If the heat loss of the matter is made up by extracting energy from the radiation, we have

$$- \Delta(a T_r^4/n m_p) = (k/m_p) \Delta T_m.$$

Therefore

$$\Delta T_m \sim - (s/k) \Delta T_r \sim - 10^8/q_0 \Delta T_r,$$

where $s/k = a T^3/nk$ is the radiation entropy per particle (§4.16) and $s \sim 10^8 k/q_0$ since $q_0 = n/n_{crit}$. In the absence of interactions we would have $T_m \ll T_r$, so $\Delta T_m \sim T_i$, the temperature at time t_i. The presence of interactions coupling the matter to radiation means that the matter temperature can be kept approximately equal to the radiation temperature, so $\Delta T_m \sim T_r$, while the latter is pulled down by a factor $\Delta T_r/T_r$, which we see is no more than about one part in $10^8/q_0$.

Whether the matter and radiation temperatures are in fact kept approximately equal depends not only on the energy balance discussed here, but on the existence of mechanisms to transfer the energy. The effectiveness of the various processes that affect this transfer determines how equal the temperatures are. Various eras in the evolution of the combined system of matter and radiation can therefore be identified in terms of the dominant interaction mechanisms. These are discussed in turn in the following sections. The presence of even only three types of particle (photons, electrons and protons) leads to numerous possibilities and concomitant confusion. The introduction of neutrinos at early times, and of neutral atoms at late times, adds to the difficulty. One has to try to remember that all one is concerned with is the elastic or inelastic scattering of photons (and neutrinos) and relativistic or non-relativistic particles in all possible combinations, most of which are irrelevant most of the time. Once a relevant interaction has been identified, we have to calculate whether the timescale over which it operates is greater or less than the timescale for the state of the universe to change by expansion. If the interaction timescale is the shorter, then as a first approximation we can assume that the interaction achieves an equilibrium at each time.

175

8.4. The Era of Equilibrium

At temperatures in excess of 10^{13} K we have $kT > 10^3$ MeV $\gtrsim m_p c^2$; this means that the average particles and photons at this stage carry sufficient energy to produce proton–antiproton, or neutron–antineutron pairs (as well as all lighter particles). An equilibrium is set up in which all such types of particles are present in numbers roughly equal to the number of photons, and with rapid interconversion of energy between the particle types. The density at this stage must be of order $(1 + z)^3 \rho_0 \sim (T/T_0)^3 \rho_0 \sim 10^8$ kg m^{-3}, which is above nuclear densities. Therefore interaction between particles via the strong nuclear force is important, and the appropriate physics is essentially unknown. In particular, we do not know the equation of state, and so we cannot predict the evolution of temperature and density before this time. Attempts have been made to discuss what might have happened, and to set limits from current observations. Suggestions include the creation of unlimited species of particles, together with a maximum limiting temperature as $t \to 0$, which leads to the equation of state (6.9.2). Another possibility is the separation of regions of matter and antimatter. This and more recent work on a 'grand unified theory' of strong, weak and electromagnetic interactions are discussed in Chapter 12. The problem is that even if one allows the belief that the microphysics is known, it is still exceedingly difficult to calculate the consequences for the macroscopic behaviour of the system. At present it is perhaps best to admit our ignorance of at least the first 10^{-5} seconds of the history of the Universe.

This would be a fatal admission if the conditions in the Universe at 10^{-5} s were determined by its earlier history, for we should not then know how to start off the remainder of the evolution that we could, in principle, predict. In the 'standard' model, with which we are concerned, this embarrassment is avoided by the assumption that we start from a state of thermal equilibrium, which, by definition, bears no memory of previous conditions. It is conceivable that the physics of the prehistory prior to 10^{-5} s does not lead to a state of equilibrium, and other models would then be possible.

As the temperature falls to somewhat over 10^{12} K there are no longer any significant numbers of pairs of heavy particles, and we are left with only the excess baryons, the postulated presence of which initially is apparently necessary to ensure that something will eventually survive (§4.16). There are still the various types of muons, and electron–positron pairs, as well as photons and the different neutrinos, all interacting through the electromagnetic and the weak nuclear forces sufficiently effectively to maintain thermal equilibrium. Since the densities involved now are well below nuclear densities, we know that the system should behave as a relativistic gas, and hence $p = \frac{1}{3}\mu$.

The muon lifetime against decay, $\mu^- \to e^- + \bar{\nu}_\mu + \gamma$, or $\mu^+ \to e^+ + \nu_\mu + \gamma$, is about 10^{-4} s, so they disappear at about 10^{12} K, their energy being shared out equally amongst the remaining species.

The participation of neutrinos in the thermal activities of the universe is also rapidly coming to an end. We can estimate the time at which this occurs by the following argument, some form of which will be the basis of much of the subsequent discussion.

In order to maintain equilibrium, a typical neutrino must interact with matter in some way after a period of freedom which is significantly less than the timescale for changes in the matter density and temperature; otherwise these changes would not be communicated to the neutrinos and equilibrium would be broken. The lifetime of a neutrino in free flight before it interacts, t_ν, will be its mean free path divided by its speed, which is c. Thus

$$t_\nu \sim \frac{1}{\sigma_\nu nc} \ll t_{\exp} \sim t,$$

where σ_ν is a typical interaction cross section, and n is the number density of obstacles. The expansion timescale, $t_{\exp}$, is taken as the age, t, of the universe at the appropriate time, this being a reasonable estimate of the time during which the expansion causes conditions to change appreciably. Equivalently, we can write the condition as $\sigma_\nu nct \gg 1$. In this form it states that the optical depth across the visible Universe, size ct at time t, is large, so neutrinos cannot travel through it unhindered. Thus we estimate $\sigma_\nu nct \sim 10$, say, as the point at which the neutrinos decouple.

We have noted that the number density of particles at this stage is about equal to the number density of photons, so $n \sim n_{ph} \sim u_{ph}/kT \sim aT^3/k$. A typical neutrino cross section is $\sigma_\nu \sim 10^{-46} \, \mathrm{m}^2$. Since the early universe is radiation-dominated we can take $t \sim 10^{20}/T^2$ (equation 6.8.8) to eliminate t.

Putting these results together, we obtain $T \sim 3 \times 10^{10}$, or $z \sim T/T_0 \sim 10^{10}$, as the stage at which the universe becomes sufficiently transparent to neutrinos that they behave independently of the remaining constituents.

We can now work out a more accurate relation between time and temperature for this equilibrium regime than the estimate given by equation (6.8.8). In the radiation-dominated models at early times we have

$$\rho \equiv \frac{u}{c^2} = \frac{u_0}{c^2}\left(\frac{a_0}{a}\right)^4 = \frac{4.5 \times 10^8}{t^2} \, \mathrm{kg \, m^{-3}} \qquad (8.4.1)$$

from equation (6.8.7). To convert this to a relation between temperature and time in a universe containing pure electromagnetic radiation we use $u = aT^3$. However, in the presence of equilibrium abundances of electron–positron pairs and neutrinos, we have instead $u = 9/2\, aT^4$.

To see how this arises recall first how the Stefan–Boltzmann law, $u = aT^4$, follows from the Planck distribution function for photons in thermal equilibrium. We have

$$u_{ph} = \frac{8\pi h}{c^3}\int_0^\infty \frac{\nu^3 \, d\nu}{\exp(h\nu/kT) - 1} = \frac{8\pi k^4}{h^3 c^3}\int_0^\infty \frac{x^3 \, dx}{e^x - 1} \cdot T^4 = aT^4$$

on introducing $x = h\nu/kT$. The numerical factor allows for two photon polarisation states.

Neutrinos differ from photons in obeying Fermi–Dirac rather than Bose–Einstein statistics with the appropriate change in the distribution function. There are also more neutrino states, since we have both electron (ν_e) and muon (ν_μ) neutrinos and their antiparticles – a total of four states. Therefore

$$u_\nu = \frac{16\pi h}{c^3} \int_0^\infty \frac{\nu^3 \, d\nu}{\exp(h\nu/kT) + 1} = \frac{16\pi k^4}{h^3 c^3} \, T^4 \int_0^\infty \frac{x^3 \, dx}{e^x + 1}.$$

In the ultra-relativistic limit ($E \gg m_0 c^2$) the electron distribution function is essentially the same as that for neutrinos; strictly, ν should be replaced by E/h, but this makes no difference to the integrals. The electron chemical potential, which might also be expected to appear, is effectively zero since electrons can be made freely from the radiation. There are four electron states, since both the positron and electron have two spin states. Therefore $u_{e\pm} = u_\nu$.

Rearranging the identity

$$\int_0^\infty \left(\frac{x^3}{e^x - 1} - \frac{x^3}{e^x + 1} \right) dx = \int_0^\infty \frac{2x^3}{e^{2x} - 1} \, dx = \frac{1}{2^3} \int_0^\infty \frac{z^3 \, dz}{e^z - 1},$$

we obtain

$$\int_0^\infty \frac{x^3}{e^x + 1} \, dx = \left(1 - \frac{1}{2^3} \right) \int_0^\infty \frac{x^3}{e^x - 1} \, dx.$$

Therefore

$$u \equiv u_{\text{Total}} = u_{\text{ph}} \left(1 + \frac{u_\nu}{u_{\text{ph}}} + \frac{u_{e\pm}}{u_{\text{ph}}} \right) = \left(1 + \tfrac{7}{4} + \tfrac{7}{4} \right) aT^4 = \tfrac{9}{2} aT^4, \qquad (8.4.2)$$

as asserted. From equation (8.4.1) and (8.4.2) we derive

$$T = 10^{10}/t^{1/2}$$

as an accurate relation, not just an order of magnitude estimate, so long as the neutrinos remain coupled to matter.

In fact, the relation holds for some time after decoupling of the neutrinos, since, even while they behave independently, the neutrinos and radiation cool in the same way, and their temperatures remain equal, $T_\nu = T_{\text{ph}} \propto (1 + z)^{-1}$. This follows because the neutrino distribution function is preserved on expansion in just the same way as the black-body spectrum. However, at $T \sim 10^{10}$ K the thermal energy per particle becomes insufficient to produce copious electron–positron pairs and this leads to a heating of the photon–particle gas, but not of the neutrinos.

We can see what happens to the particle pairs existing at 10^{10} K using the standard argument. The cross section for annihilation of $e^\pm$ pairs is of order the Thomson cross section, $\sigma_T \sim 10^{-28} \, \text{m}^2$. The time that an average electron must wait before

178

annihilation is $t_{e^\pm} \sim (\sigma_T n_e c)^{-1}$, where n_e is the electron density. This follows either because the mean free path, $c t_{e^\pm}$ is $(\sigma_T n_e)^{-1}$, or because $\sigma_T n_e c$ is the rate of reaction per incident particle travelling at a speed approximately equal to c at a target density n_e. The electron density is $n_e \sim n_{ph} \sim (aT^3/k) \sim 10^{37} \, \text{m}^{-3}$ at $T \sim 10^{10} \text{K}$, so $t_{e^\pm} \sim 10^{-17} \text{s}$. At 10^{10}K, the age of the universe is 1 s, and, as before, this is a reasonable estimate also of the expansion timescale, t_{exp}. Therefore $t_{e^\pm} \ll t_{exp}$, and the electron–positron pairs have ample time to annihilate before anything else happens.

However, the annihilation energy cannot now be shared out to the neutrinos, since they are essentially non-interacting. It therefore goes almost entirely into photons, with also a negligible addition to the thermal energy of matter, so after annihilation we must have $T_{ph} > T_\nu$. Our estimate of u (equation 8.4.2) is therefore no longer valid.

To calculate the correct relation assume, justifiably, that after each increment in annihilation energy the electron–photon gas has time to come into equilibrium at its new temperature. The system then moves through a sequence of equilibrium configurations and we can apply the usual laws of equilibrium thermodynamics.

At each step the heat energy, dQ, extracted from the $e^\pm$ annihilation, is supplied to the photon gas at the same temperature. Therefore

$$dQ = -T \, dS_{e^\pm} = T \, dS_{ph},$$

where $S_{e^\pm}$ and S_{ph} are the entropies of electron–positron pairs and photons, respectively, in a given region of space of volume V. Therefore the entropy of the pairs goes into photon entropy. Before annihilation starts we have $T_{ph} = T_{e^\pm} = T_\nu$, and

$$s_{ph+e^\pm} = \frac{S_{ph}}{V} + \frac{S_{e^\pm}}{V} = \tfrac{4}{3}aT^3_{ph} + \tfrac{7}{3}aT^3_{e^\pm} = \tfrac{4}{3}(1+\tfrac{7}{4})aT^3_{ph} = \tfrac{11}{4} s_{ph},$$

where the entropy density of the photons, $s_{ph} = S_{ph}/V$, and of the electron–positron pairs, $s_{e^\pm} = S_{e^\pm}/V$, are calculated as in §3.6 using the energy densities from equation (8.4.2). Similarly, $s_\nu = \tfrac{7}{4}s_{ph}$, and therefore

$$s_{ph+e^\pm}/s_\nu = \tfrac{11}{7}.$$

This ratio is constant throughout the pre-annihilation period since both numerator and denominator are proportional to T^3_{ph}. The key to the argument is that, since all the entropy of the pairs goes into photons, this ratio must also equal the ratio s_{ph}/s_ν after annihilation is complete. At that stage we have $T_{ph} \neq T_\nu$, so

$$s_{ph}/s_\nu = \tfrac{4}{3}aT^3_{ph}/\tfrac{4}{3}aT^3_\nu = \tfrac{11}{7}.$$

Therefore

$$T_{ph} = (\tfrac{11}{4})^{1/3} T_\nu = 1.4 T_\nu,$$

and the photon temperature is increased by 40% by the annihilation.

The total energy density after annihilation is

$$u = u_{ph} + u_\nu = u_{ph} \left[1 + \tfrac{7}{4} \left(\frac{T}{T_{ph}} \right)^4 \right] \sim 1.45 \, u_{ph},$$

and from this and equation (8.4.1) we obtain

$$T_{ph} = \frac{1.4 \times 10^{10}}{t^{1/2}} \, K,$$

with t in seconds. This confirms that our estimate (equation 6.8.8) is reasonable for order of magnitude calculations, and so our discussion up to this point is consistent!

8.5. The Era of Nucleosynthesis

A standard universe at about 1 s contains photons, electrons, protons and neutrons in thermal equilibrium at 10^{10} K, as well as neutrinos at a somewhat lower temperature. The neutrinos are thermally decoupled, but they still interact inelastically with neutrons and protons to a significant extent converting between the two nucleon states according to the reactions:

$$p + \bar{\nu}_e \rightarrow n + e^+$$

$$n + \nu_e \rightarrow p + e^-.$$

The point is that these reactions involve very few neutrinos, and so are of no consequence for the neutrino distribution. However, they involve significant numbers of the far less numerous protons and neutrons, and so cannot be neglected in the distribution of nucleons.

In the standard model, and in our subsequent discussion, it is assumed that the abundances of neutrinos and antineutrinos are equal. A sufficient overabundance of one form or the other would lead to an imbalance between protons and neutrons that would affect the synthesis of helium to the extent that it could be produced in arbitrary amounts. We shall not consider this. In principle, the neutrino pairs can annihilate into photons, but the interaction is so weak that the annihilation time is very much longer than the age of the Universe, and annihilations can be neglected.

The efficient interconversion of neutrons and protons keeps these at their statistical equilibrium abundances, given approximately by the Boltzmann formula

$$n_n/n_p = \exp \left[-(m_n - m_p) \, c^2/kT \right],$$

where n_n and n_p are the number densities, and m_n and m_p are the masses of neutrons and protons. In this approximation we neglect the statistical weights of the free electrons and neutrinos. The interconversion effectively ceases at about 10^9 K, because the reaction rates become less than the expansion rate of the universe. The universe therefore enters its next phase with the neutron-proton ratio fixed, or 'frozen-in', at its value at 10^9 K. If we approximate this switching off of the

180

reactions as a sudden step, and use the Boltzmann formula, the frozen-in ratio is

$$f \equiv (n_n/n_p)_{T=10^9 K} = \exp\left[-(m_n - m_p) c^2/10^9 k\right] \approx 0.2.$$

A more accurate computation including the gradual switching off of the reaction gives

$$f \approx 0.15.$$

At all stages in the evolution so far, protons and neutrons have been reacting to form deuterium:

$$n + p \rightarrow d + \gamma. \qquad (8.5.1)$$

However, deuterium is very easily destroyed by the reverse reaction, and above about $10^9 K$ it is broken up almost as soon as it is formed. The concentration of deuterium at this stage is therefore negligible. Below $10^9 K$ however, the deuterium abundance starts to build up to the extent that its reaction with a further proton becomes important:

$$d + p \rightarrow {}^3He + \gamma;$$

^{3}He then begins to accumulate until it starts to produce ^{4}He at a significant rate:

$${}^3He + {}^3He \rightarrow {}^4He + 2p.$$

If the density is very low, the timescale for this synthesis will be more than the expansion time, and helium will not be produced abundantly. If the density is too high, the helium will react further, to yield first ^{12}C, via the 'triple-α' reaction ($3\,{}^4He \rightarrow {}^{12}C$), and then heavier elements by further reactions. In the intermediate case all the neutrons initially present will be incorporated into ^{4}He, so we can calculate the expected helium abundance.

The proportion of ^{4}He by number is,

$$\frac{n_{^4He}}{n_p + n_{^4He}} = \frac{\tfrac{1}{2}n_n}{n_p + \tfrac{1}{2}n_n} = \frac{f}{2+f} \sim 7\%$$

and the ratio by mass is

$$Y = \frac{m_{^4He} \cdot n_{^4He}}{m_p n_p + m_{^4He} n_{^4He}} \approx \frac{2f}{1 + 2f} \approx 25\%.$$

This is close to the observed cosmic abundance of helium. We can estimate the conditions at $\sim 10^9 K$ required to yield this abundance by a simple argument; the results of the subsequent discussion are confirmed by detailed numerical integration using the correct reaction rates for the hundred or so nuclear reactions of potential importance.

In order that the reactions should operate to produce some helium, the reacting particles must not be pulled apart significantly before they have had time to react. Therefore, the reaction time must be less than the expansion time at $T \sim 10^9 K$, when the reaction (8.5.1) first starts to become important. If the cross section for the slowest reaction in the chain is σ_{He}, the reaction rate per target particle is $\sigma_{He} nv$ for a nucleon density n and velocity v, since a flux nv particles per second per unit area is incident on the target. Therefore we require

$$(\sigma_{He} nv)^{-1} < t_{exp} \sim t,$$

where, as usual, we estimate the expansion timescale as the age of the universe at the appropriate time. Then $v \sim (3kT/m_p)^{1/2} \sim 3 \times 10^6\,\text{m s}^{-1}$ is a typical velocity, and $t \sim 10^{20}/T^2 \sim 10^2\,\text{s}$. This yields

$$n \gtrsim 3 \times 10^{-9} \sigma_{He}^{-1}$$

for the nucleon number density at $10^9\,\text{K}$. We therefore estimate

$$s = \frac{4}{3}\frac{aT^3}{n} \lesssim 10^{43}\, \sigma_{He}\, k.$$

Note that the typical particle velocities are well within those attainable in the laboratory, so the cross sections are known with reasonable accuracy. We can take $\sigma_{He} \sim 10^{-33}\,\text{m}^2$ as appropriate, giving $n \gtrsim 10^{24}\,\text{m}^{-3}$ and

$$s \lesssim 10^{10} k.$$

The particle density is somewhat less than that of air, so the conditions in the universe at the time of helium formation are by no means exotic, and laboratory physics can be applied with confidence.

If the density is too high, helium will be converted into heavier elements. However, the cross section $\sigma_{3\alpha}$ for the triple-α reaction is several orders of magnitude smaller than that for deuterium formation, so the upper limit on n is several orders of magnitude greater than the lower limit, and this does not represent a severe constraint.

On the other hand, $T \propto s^{1/3}$, so the temperature of the microwave background is determined to within an order of magnitude. Since s is constant in time, we have

$$T_0/10^9 \lesssim (n_0/10^{24})^{1/3}$$

For a present density $m_p n_0 \sim 10^{-27}\,\text{kg m}^{-3}$, we obtain $T_0 \lesssim 10\,\text{K}$, in good agreement with observation. Therefore, the big-bang nucleosynthesis model not only yields the observed helium abundance for a wide range of initial conditions, but gives also the microwave background temperature correctly. In fact, it was this argument that led Gamow and Dicke, independently, to *predict* the existence of a microwave background.

The observed abundance of deuterium can now be used to provide more stringent limits on the density of matter. The point is that deuterium is so relatively easy to destroy that it is difficult to suggest any site for its synthesis, other than in the early universe. We therefore expect the observed abundance to represent a lower limit to cosmologically synthesised deuterium. Even in the early universe, conditions must be just right if a significant amount of the deuterium is to remain unreacted. In figure 8.3 the results of numerical computations by Wagoner (1974) are displayed, which show that the observed abundance of deuterium is produced in a universe with $s \sim 5 \times 10^9 k$.

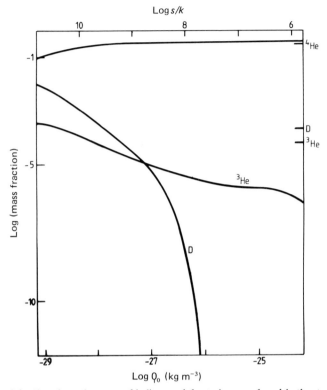

Figure 8.3. Abundance by mass of helium and deuterium produced in the standard hot big-bang models for various values of the present density or entropy per baryon. Solar system abundances are shown on the right of the graph (adapted from Wagoner 1974).

To understand how this comes about, suppose that the effective cross section for the destruction of deuterium is σ_D. Synthesis of deuterium must be significantly reduced after $t \sim 700$ s, because this is the half-life of a free neutron, and so by this time a typical unreacted neutron will have beta-decayed. Therefore, if significant deuterium is to remain, we require that at this time the timescale for its destruction should be *greater* than the expansion timescale. Hence we require

$$(\sigma_D nv)^{-1} > t_{exp}$$

at $t \sim 700$ s, or $T \sim 4 \times 10^8$ K. This gives, by the same argument as before,

$$s \gtrsim 10^{42} \sigma_D k.$$

183

For the sake of discussion we can take $\sigma_D \sim \sigma_{He}$ for the cross sections at the appropriate energies, from which we obtain strong limits on s, namely $10^9 k < s < 10^{10} k$, and hence on q_0. The numerical computations give $0.02 < q_0 < 0.1$ if we are to obtain the observed deuterium abundance (Gott et al 1974). Elements heavier than helium and deuterium can be made and returned to the interstellar medium by processes other than cosmological synthesis, so the argument cannot be extended.

We conclude that, provided the deuterium measurements have been interpreted correctly, and if all the deuterium is cosmological, the matter density in the universe can be no more than a few per cent of the critical density. This provides strong, but not invincible, indirect evidence for an open Universe.

8.6. The Plasma Period

The Universe at around $5 \times 10^8 K$ has become a radiation-dominated plasma of hydrogen and helium with a matter density somewhat less than that of air. For simplicity we shall henceforth ignore the presence of nuclei other than hydrogen. Nevertheless, we should check that the production of nuclear energy during nucleosynthesis does not distort the Planck spectrum of the radiation. The worst possible assumption, which in practice is far from attainable, is that all the protons are converted to helium and all the nuclear energy released goes into radiation. For each proton we get an energy of about $10^{-2} m_p c^2$, so, as a fraction of the background energy we get an input of

$$\frac{10^{-2} m_p c^2 n_p}{a T^4} \sim 10^{-5}.$$

This is entirely negligible, so the spectrum of radiation remains black-body. In fact, this input of energy is rapidly thermalised, and raises the temperature by a tiny amount.

Protons and electrons are strongly coupled together at all times by elastic scattering, and this keeps them at the same temperature. Photons are coupled to matter at first by free–free absorption and emission. At early times the plasma is optically thick to true absorption and remains in thermal equilibrium. However, at the low densities and relatively high temperatures encountered at the end of nucleosynthesis, the free-free opacity is already rather small, and decreasing with the expansion. At a frequency $\nu \sim kT/h$, around the peak of the Planck curve, the absorption coefficient is

$$\mu_{ff} \sim 10^{-28} n^2 T^{-7/2} \sim 10^{-30} T^{5/2} \ (m^{-1})$$

if $n = n_{crit}$. The final expression follows since $n \propto (1+z)^3 \propto T^3$. As usual, we have used the marginally unbound model, corresponding to an Einstein–de Sitter model in the late stages of evolution, to provide a numerical example. Thus the absorption time is

$$t_{ff} = (\mu_{ff} c)^{-1} \sim 10^{22} T^{-5/2} s.$$

184

In order to estimate the expansion timescale, we assume the universe to be radiation-dominated when it becomes thin to free–free absorption, and check this presently. Therefore the expansion timescale is $t_{exp} \sim t \approx 10^{20}/T^2$. The universe is optically thin when $t_{ff} > t$, which occurs for $T \lesssim 10^8 \text{K}$. This is consistent with our assumption that it is radiation-dominated at the transition (see §6.8).

Nevertheless, it does not follow that equilibrium between matter and radiation is broken at 10^8K. The reason is that there are further processes which couple the radiation and matter together. In particular, the scattering of photons by electrons is still important, and becomes dominant with the demise of free–free transitions. (The cross section for scattering by protons is a factor $(m_e/m_{p\delta})^2$ smaller and so is always negligible.) We can confirm this easily: the electron scattering timescale is

$$t_T = (\sigma_T n_e c)^{-1} \sim 25(1+z)^{-3}$$

for $q_0 = \frac{1}{2}$. Here the self-consistent assumption will turn out to be that the universe is matter-dominated at the stage when $t_T \sim t_{exp}$. Hence the appropriate estimate of the expansion timescale is

$$t_{exp} \sim t \sim 1/H = (1+z)^{3/2}/H_0,$$

rather than (6.8.8). This gives $t_T \ll t_{exp}$ as long as $z > 10^3$, at which point the universe is matter-dominated, as assumed. In fact, as we have noted in § 8.3, other considerations become relevant before we reach $z = 10^3$.

However, Thomson scattering is not a true absorption process; there is no energy exchange between photons and electrons. If Thomson scattering alone were important then in a homogeneous isotropic universe only the identity of the photons would be changed by scattering, and not the overall distribution of energy. Therefore the matter and radiation would cool independently.

The actual situation is more complex. We shall deal in a moment with the fact that Thomson scattering is only an approximation in need of detailed amendment. First, however, note that the effect of Thomson scattering is to keep the photons shuttling around a more restricted part of the universe than they would otherwise have been able to visit in a given time. For the purpose of getting from A to B, the photons are effectively travelling more slowly than they would in the absence of scattering. We can work out how much more slowly by approximating the journey as a one-dimensional random walk (although, in fact, the probabilities of scattering into all angles are not equal).

After a large number of steps, n, of its random walk, the photon will have travelled a distance $n^{1/2}\lambda_T$, where λ_T is the mean free path to Thomson scattering. Let λ_{ff} ($\gg \lambda_T$) be the mean free path for free–free absorption. The number of scatterings required to travel this length of path is $n = \lambda_{ff}/\lambda_T$. After this many scatterings the average photon is absorbed and has travelled a distance from its starting point $\bar{\lambda} = n^{1/2}\lambda_T = (\lambda_{ff}\lambda_T)^{1/2}$. This is therefore the effective mean free path for absorption.

The effective optical depth to true absorption is therefore $(\tau_{es}\tau_{ff})^{1/2} = (\mu_{es}\mu_{ff})^{1/2}ct$. To estimate the age of the universe we take the Einstein–de Sitter model, since the self-consistent assumption will turn out to be that the universe is still thick in the matter-dominated era. Thus $t = H_0^{-1}(1+z)^{3/2}$, and $(\mu_{es}\mu_{ff})^{1/2} \sim 10^{-21}(1+z)^{11/4}$ at the peak of the Planck spectrum, so we have an optical depth of 10 down to $z \sim 3 \times 10^3$, or $T \sim 10^4\,\mathrm{K}$.

In fact, for most of the range $3 \times 10^4\,\mathrm{K} < T < 10^8\,\mathrm{K}$, the electron temperature is held at the radiation temperature by a different mechanism. In scattering off of an electron of velocity v, a photon undergoes a finite change in frequency given by $\delta\nu/\nu \sim (v/c)^2$. Thomson scattering is correct to order v/c. If the change in frequency is important we refer to Compton scattering. Note, however, that we are referring to one and the same process considered in different regimes and to different orders of accuracy.

In a thermal distribution of electrons with velocities v, the average of v^2 is of order kT/m_e. Each scattering by an electron shifts the photon frequency on average by $\delta\nu/\nu \sim kT/m_e c^2$. The number of scatterings suffered in random walking across a distance ct_C is given by $N^{1/2}\lambda_T = ct_C$, where λ_T is again the Thomson scattering mean free path. Note that the probability of scattering is still governed to a sufficient accuracy by the Thomson cross section. Since the photon performs a random walk in frequency space, with step length $\delta\nu/\nu \sim kT/m_e c^2$, after N steps we expect to find it at $\Delta\nu/\nu = N^{1/2}(kT/m_e c^2) = (ct_C/\lambda_T)(kT/m_e c^2)$. The process is certainly important on a timescale, t_C, for which $\Delta\nu/\nu \sim 1$. This gives the Compton timescale,

$$t_C \sim \frac{1}{\sigma_T n_e c}\left(\frac{m_e c^2}{kT}\right).$$

We find that this is less than t_{exp}, and hence that the frequency change is significant, if $z \gtrsim 3 \times 10^4$ in the Einstein–de Sitter model.

To see which of the two processes dominates we compare the effective free–free timescale, $t_{esff} \sim c^{-1}(\mu_{es}\mu_{ff})^{-1/2} \sim 10^{19}(1+z)^{-11/4}$, with the Compton timescale $t_C \sim 10^{28}(1+z)^4$. Clearly, $t_C < t_{esff}$ for $z \lesssim 10^8$, so Compton scattering dominates the coupling between radiation and matter, and is responsible for the transfer of energy which stops the matter from cooling faster than the radiation.

This does not, of itself, imply that we shall have a Planck spectrum for the radiation. There could be a large distortion of the spectrum even though there is no significant net extraction of energy. In fact, in general, electron scattering *cannot* produce a thermal spectrum because it conserves the number of photons. In a thermal spectrum the number of photons is governed by the temperature, and is not a free parameter. So, if we start off with too few photons and then let only electron scattering operate, we shall end up with too few photons for a black-body spectrum; and if we start off with too many, we shall end up with too many. If, however, we start off with a thermal distribution, then to the extent that the argument of §8.3 applies, and the temperature is reduced significantly only by expansion and not by energy loss to the matter, we shall continue to have the right number of photons to maintain a thermal distribution. This is confirmed by a more complete theoretical treatment, in which the kinetic equation for photons is modified to include a collision term arising from scattering off of thermal electrons.

In fact, even this account is somewhat oversimplified. The full discussion shows that the thermal distribution is maintained to a good accuracy even with a modest energy loss to matter. The point is that the loss of energy leaves the radiation with too many photons for its energy, relative to a black-body distribution, but these excess photons are Compton scattered to low frequencies where they do not significantly distort the spectrum.

There is one further complication that should be mentioned here to avoid confusion. The condition for Compton scattering to maintain thermal coupling between matter and radiation is that the scattering timescale be significantly less than the time taken for a photon to cross the universe at the appropriate epoch, or, equivalently, that the optical depth satisfy $\tau(kT/m_e c^2) \gtrsim 1$. Consider now another case in which radiation and matter are at different temperatures and the condition that the matter does not significantly distort the radiation spectrum is required. This is relevant if, for example, the intergalactic medium has been reheated and hence reionised at some late stage. In this case we consider what happens not to an average photon, to which t_C refers, but in the worst case. After N steps we could find a photon with $\Delta \nu/\nu \sim N(kT/m_e c^2)$. Since the optical depth to electron scattering across the system is related to the number of scatterings by $\tau = N^{1/2}$, distortions to the spectrum will occur if $\tau^2(kT/m_e c^2) \gtrsim 1$. In general the relevant quantity to be compared with the optical depth is the larger of $(m_e c^2/kT)$ and $(m_e c^2/kT)^{1/2}$.

Below $T \sim 10^4$ K $(z \sim 3 \times 10^3)$, the photons are not significantly affected by the presence of the plasma since the relevant optical depths are small. However, this does not mean that the matter and radiation behave independently at this stage. To see this, note that the optical depths tell us what happens to the average photon but not how the average electron behaves. The relatively few electrons can, and do, make many collisions with photons even though most of the photons do not collide at all. For a quantitative discussion of this point the quantity to calculate is the mean free path of the *electrons* to Thomson scattering, which is $(\sigma_T n_{ph} c)^{-1}$. The electron Compton scattering timescale is therefore

$$t_{ce} = (n_e/n_{ph}) t_C,$$

and this is the time needed for an electron to gain a significant energy from the photons. Because $n_{ph} \gg n_e$, we have $t_{ce} \ll t_{exp}$ right down to $z \sim 10^3$. Of course, the net energy transfer must cease as soon as the average electron and photon energies are equal, so we conclude that Compton scattering keeps the electron temperature up to the radiation temperature down to $z \sim 10^3$, at which point the universe enters a new phase.

187

8.7. The Recombination Era

As the plasma cools sufficiently electrons and alpha particles start to recombine to helium atoms, and, at somewhat lower temperatures, protons recombine to neutral hydrogen. If the system is in thermal equilibrium, with radiation and matter at the same temperature, or if the dominant ionisation mechanism is the collision of atoms and electrons having Maxwellian distributions, rather than photoionisation, then the Saha equation gives the equilibrium degree of ionisation. In fact, at recombination, neither of these conditions is a particularly good approximation: the matter and radiation are starting to decouple and therefore to cool at different rates, and the radiation plays an important role in maintaining the ionisation equilibrium. Nevertheless, we can use the Saha equation to give an order of magnitude estimate for the epoch of recombination.

We consider only hydrogen since this is the dominant species. If $x = n_e/n$ is the ratio of electron number density to the total density ($\sim$ proton density) at temperature T, the Saha equation reads

$$\frac{x^2}{1-x} = \left(\frac{2\pi m_e k}{h^2}\right)^{3/2} \exp(-I_H/kT)\, T^{3/2} n^{-1} \sim 10^{21}(1+z)^{-3/2} \left(\frac{T_0^{3/2}}{n_0}\right) \exp\left(\frac{1.6 \times 10^5}{(1+z)\,T_0}\right),$$

(8.7.1)

where $I_H (= 13.6\,\text{eV})$ is the ionisation potential of hydrogen. Let us assume that $x \approx 1$, and estimate the redshift at which this assumption fails. For $T_0 = 3\,\text{K}$, we have

$$-\ln(1-x) \sim 40 - \tfrac{3}{2}\ln(1+z) - 5 \times 10^4/(1+z), \qquad (8.7.2)$$

for the Einstein–de Sitter model, but with only a negligible dependence on q_0. If $1 - x \sim 0$ the left-hand side of (8.7.2) is large and positive, and this ceases to be consistent with the right-hand side at redshift somewhat larger than 10^3. Thus the equilibrium favours recombination at temperatures below about $3 \times 10^3\,\text{K}$.

Of course, as always, we should check that equilibrium can be achieved despite the expansion of the universe, by comparing the recombination and expansion timescales. The rate at which electrons and protons recombine is given by $\sigma_R v n_e n_p$ per unit volume or, since $n_e = n_p \sim n$ initially, by $\sigma_R v n_e$ per particle. Here σ_R is the recombination cross section for an electron with velocity v. Averages of the product $\sigma_R v$ for a thermal plasma are tabulated as recombination coefficients $\alpha(T)$ (e.g. Allen 1972). With $\alpha(T) \sim 10^{-16}\, T^{-1/2}\,\text{m}^3\,\text{s}^{-1}$ we find the recombination time $t_R = (\alpha n_e)^{-1} \ll t_{exp}$ for z near 10^3. Left to itself, without any external energy input, the plasma recombines. From this point to the present the matter and radiation cool independently.

Finally, we must investigate whether the energy released by recombination can distort the Planck spectrum of the background radiation. For each recombination the energy released is $I_H \sim 13.6$ eV; there are of order n recombinations per unit volume at temperature T, so the ratio of ionisation energy to background radiation is

$$\frac{I_H n}{aT^4} \sim \frac{I_H n_0}{aT_0^4(1+z_R)} \sim 10^{-6}.$$

This energy is emitted in the ultraviolet and optical and redshifted to the far-infrared where it represents a negligible addition to the expected emission from Galactic dust.

We have therefore confirmed that an initial Planck spectrum is essentially unperturbed in the expansion of the Universe, despite the presence of matter. This then is the canonical hot big-bang model. In Chapter 9 we shall turn the argument round and ask how different from this picture we could allow the Universe to have been in the past, and yet still ensure that we end up with something sufficiently close to present observations.

8.8. Last Scattering

The background radiation is a signal from creation, bringing us information from the earliest phases of the Universe. Nevertheless, when we look out at the microwave sky we do not see the creation itself, just as when we look at the Sun we do not see directly to its central nuclear fire. To a first approximation what we see is a surface at unit optical depth away from us. In the case of the Universe the only significant opacity at recent times is provided by electron scattering, so the microwave background appears to come to us from the surface where the radiation was last scattered. It is of interest to calculate the redshift of this surface.

This calculation is very different from those we have pursued up to now in this chapter. Previously we have asked what would happen to radiation in a universe, the expansion of which was frozen at a given epoch. Now we are asking how radiation is transmitted through the expanding universe. To compute this we must solve the radiative transfer equation including scattering. In fact, in this case the solution is almost immediate once the equation has been written down.

In general, the intensity of radiation changes along a ray not only as a consequence of expansion, but also from emission and scattering processes. The additional change in intensity $di_{\nu(1+z)}$ in time dt, at a point along the ray having redshift z relative to the observer, is equal to the emission per unit area into unit solid angle of the beam, less the loss from the beam, and is given by

$$j_{\nu(1+z)}(t) \, c \, dt - \sigma_{\nu(1+z)} n(t) \, i_{\nu(1+z)}(t) \, c \, dt.$$

The factors of c arise from the conventional choice of units for the emission coefficient j_ν, and the absorption cross section σ_ν. The emission coefficient includes scattering into the beam, and the absorption is taken to include scattering out of the beam.

189

Using the result of §8.1 (in particular, equation 8.12), the full transfer equation is now

$$\frac{di_{\nu(1+z)}(t)}{ds} - \frac{3}{(1+z)} i_{\nu(1+z)} \frac{dz}{ds} = j_{\nu(1+z)} - \sigma_{\nu(1+z)} n(t) i_{\nu(1+z)}(t), \quad (8.8.1)$$

where the parameter s is defined here to be $s \equiv ct$. For the computation of the last-scattering surface, we can neglect scattering and emission into the beam, so $j_\nu = 0$. We define the optical depth at the *observed* frequency ν by

$$\tau_\nu = \int \sigma_{\nu(1+z)} n(t) c \, dt.$$

where z is, of course, a function of time, t. For the present case, the cross section is independent of frequency and (8.8.1) becomes

$$\frac{di_{\nu(1+z)}}{d\tau} - \frac{3}{(1+z)} i_{\nu(1+z)} \frac{dz}{d\tau} = - i_{\nu(1+z)}.$$

The solution for the intensity observed at frequency ν is

$$i_\nu(0) = \exp(-\tau_*) \frac{i_\nu(1+z_*)}{(1+z_*)^3} \,,$$

where $i_{\nu(1+z_*)}$ is the radiation emitted at some initial redshift z_*, and

$$\tau_* = - \int_0^{z_*} \sigma_T n_e c \, dt.$$

We have therefore obtained an expression for the appropriate optical depth representing the attenuation of an initial input spectrum in the usual way.

If n_e is given, we can proceed to calculate the redshift at which $\tau_* = 1$. In the canonical model we can assume n_e to be given by the Saha equation (8.7.1), but we must now make the approximation $x \equiv n_e/n \ll 1$. This gives

$$n_e = \left(\frac{2\pi m_e k}{h^2}\right)^{3/4} n_0^{1/2} T_0^{3/4} (1+z)^{9/4} \exp\left(-\frac{I_H}{2kT_0} \cdot \frac{1}{1+z}\right).$$

Again, as an example, consider the Einstein–de Sitter model. Then

$$\tau_* = - \int_0^{z_*} \sigma_T n_e c \frac{dt}{dz} dz = \int_1^{1+z_*} \sigma_T n_e c H_0^{-1} (1+z)^{-5/2} d(1+z),$$

and therefore, approximately,

$$\tau_* \approx 10^8 \int_1^{1+z_*} (1+z)^{-1/4} \exp\left(-\frac{2.7 \times 10^4}{1+z}\right) d(1+z).$$

To estimate the integral assume that $1 + z_* \ll 2.7 \times 10^4$. Then

$$\tau_* \lesssim 10^8 (1 + z_*)^{3/4} \exp\left(-\frac{2.7 \times 10^4}{1 + z_*}\right).$$

At decoupling, $z \sim 10^3$, this yields $\tau_* \lesssim 1$. However, by this stage the assumption that the gas is only slightly ionised, $x \ll 1$, is no longer valid. As soon as the ionisation becomes significant near $z \sim 10^3$ the optical depth rises very sharply. So we conclude that in the high-density ($q_0 = \frac{1}{2}$) case with no reionisation of the intergalactic medium we would see right back to the time of decoupling. In fact, the result is only weakly dependent on the density.

However, we know that a high-density, recombined intergalactic medium would conflict with the Lyman-α observations discussed in §4.10. Therefore, while we might have $z_* \sim 1000$ in a low-density universe, a more realistic case to consider in the contest of a high intergalactic density is a plasma fully ionised by reheating. Then $n_e = n_0 (1 + z)^3$ and we obtain for τ_*

$$\tau_* = \frac{\sigma_T n_0 c}{H_0} \int_0^{z_*} (1 + z)^{1/2} dz \sim 0.04 [(1 + z_*)^{3/2} - 1].$$

This yields $\tau_* = 1$ at $z \sim 7$ as the surface of last scattering. In this case the observed isotropy of the microwave background translates into a remarkable degree of isotropy of the Universe at a redshift of 7, corresponding to an age of about 5×10^8 years.

8.9. Horizons and Causality

In the Newtonian picture of a universe of finite age t, we expect to be able to see to a distance ct at most, for from beyond this there has been no time for signals to reach us. Consequently, we expect there to exist a boundary to the visible Universe, or horizon, at any time. It is easy to confirm this expectation for the particular general relativistic cosmological models of Chapter 6, although other models are possible which do not have horizons (see §10.3).

Take the dust models as examples. Let an observer at the origin of Robertson-Walker coordinates (equation 6.4.7) at time t_0 look back to galaxies of redshift z. The radial coordinate R of such galaxies is provided by equation (7.2.5) which gives the time (or redshift) at which light must be emitted by a galaxy at $\chi = \chi_e$ in order to be seen at $\chi = 0$ at $t = t_0$. Thus

$$\frac{a_0 R}{c} = \frac{a_0 \sin \sqrt{k}\, \chi_e}{c \sqrt{k}} = \frac{q_0 z + (q_0 - 1)[(2 q_0 z + 1)^{1/2} - 1]}{q_0^2 (1 + z) H_0}.$$

As $z \to \infty$, we see that

$$a_0 R \to a_0 R_\infty = c (H_0 q_0)^{-1}. \tag{8.9.1}$$

If $0 < q_0 \leqslant \frac{1}{2}$ the whole Universe is covered by $0 < R < \infty$. Since, according to equation (8.9.1), R_∞ remains finite as $z \to \infty$, the galaxies we can see do not constitute the whole Universe, and we have established the existence of a horizon in these models.

The argument does not work so easily for $q_0 > \frac{1}{2}$ since $|R| \leqslant 1$ here anyway. Indeed, horizons need not exist at sufficiently late times in these models, when photons have been able to cross the Universe. However, we can readily show that closed dust models have horizons at early times. In the coordinates of equation (6.4.8), a radially directed null ray satisfies

$$d\chi = \pm \frac{dt}{a(t)}.$$

Therefore, the range of χ-coordinates covered by a light ray between $t = 0$ and $t = t_0$ is

$$\Delta\chi = \int_0^{t_0} \frac{dt}{a(t)} = \int d\phi.$$

The last equality follows on introducing the ϕ-coordinate of equations (6.7.14) and (6.7.11). For $t < t_{max}$, we have $\phi < \pi$, so $\Delta\chi < \pi$. This means that a photon crosses less than half the Universe and for times prior to the point of maximum expansion there exist particle horizons (figure 8.4).

In the dust model a photon just has time to travel around the Universe between the initial and final singularity. This is a special feature of the dust model. A similar diagram of the radiation-dominated model shows that a photon can traverse no more than one-half of the Universe, and a particle horizon exists for all observers outside the final singularity (figure 8.5).

It is clear from the diagrams that the horizon increases with time as more and more galaxies come into view. Analytically, this follows by computing dR_∞/dt_0. For example, for $q_0 = \frac{1}{2}$ we have from equation (8.9.1)

$$R_\infty = 2(a_0 H_0)^{-1} c = 2\dot{a}_0^{-1}$$

and hence

$$\dot{R}_\infty = a^{-1} > 0.$$

Similarly, the redshift of a given galaxy decreases with time. For $q_0 = \frac{1}{2}$, equation (7.2.5) gives

$$a_0 \chi = \frac{2c}{H_0} [1 - (1+z)^{-1/2}].$$

A given galaxy has $\chi =$ constant, so for small z, or, equivalently, $t_0 \to \infty$, we have

$$z(t_0) \propto a_0 H_0 \propto t_0^{-1/3},$$

which is a decreasing function of time.

192

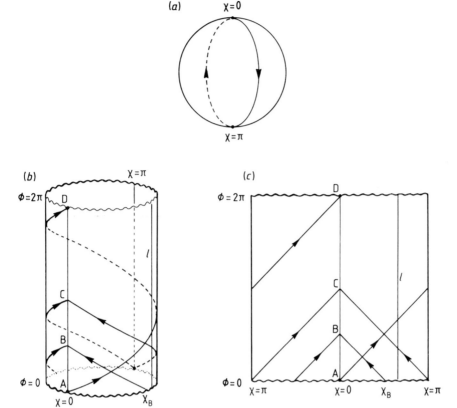

Figure 8.4. (*a*) A point travelling round the surface of a sphere moves from polar angle $\chi = 0$ at the north pole to $\chi = \pi$ at the south pole, and then back again from $\chi = \pi$ to $\chi = 0$. (*b*) The (χ–ϕ) coordinate plane of the $k = +1$ Robertson–Walker dust models, represented on the surface of a cylinder. A photon emitted at the initial singularity at A goes once around the universe before encountering the final singularity at D. An observer at B can see galaxies within his horizon, $\chi = \chi_B$. An observer at C can see the whole universe. The line l is the world-line of a galaxy that cannot be seen at B. (*c*) The cylinder of (*b*) unwrapped to a plane. Note that the representation of the coordinate plane does not illustrate proper distances; the universe has zero volume at $\phi = 0$ and $\phi = 2\pi$, and its maximum volume at $\phi = \pi$.

The existence of horizons presents a profound source of puzzlement in our isotropic Universe. To understand why one should be puzzled, note first that horizons divide the world into regions which cannot have communicated with each other by a given cosmic time, and so cannot have influenced each other. Now consider two microwave photons propagating to us from directions $\Delta\theta$ apart in the sky. These photons last interacted with matter at the last-scattering surface. This is at $z_* \sim 7$ in the reheated Einstein–de Sitter model, which we can consider for the sake of argument. More remote last-scattering surfaces do not alter the argument in any important way. Let the photons come from points A and B in figure 8.6. We can ask whether the conditions at A can in any way have been influenced by those at

193

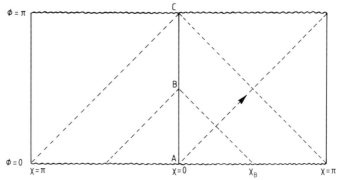

Figure 8.5. (χ–ϕ) plane of the radiation-dominated Robertson–Walker model ($k = +1$). A photon emitted at A crosses half the universe before being engulfed in the final recollapse. An observer at B has a horizon at χ_B. An observer at C would be able to see the whole universe were he not crushed in the final singularity at this point.

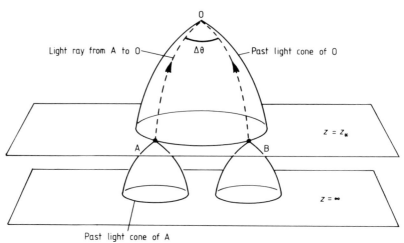

Figure 8.6. An observer at O sees two points A and B on the last-scattering surface an angle $\Delta\theta$ apart. These points cannot have been in causal contact at z_* since their past light cones do not intersect.

B. They cannot have been if B is outside the horizon of A at z_*. From equation (8.9.1) the condition for this is that the separation of A and B should exceed

$$a_* R_\infty = \frac{2c}{H_*},$$

where starred quantities are evaluated at z_*. The angle subtended at the observer by two points this proper distance apart at z_* is, from equation (7.2.6) with $q_0 = \frac{1}{2}$,

$$\Delta\theta = \frac{(1+z)^{1/2}}{z - (1+z)^{1/2} - 1}.$$

This gives $\Delta\theta \sim 30°$.

194

Now consider microwave radiation from two directions greater than $30°$ apart. Even without allowing for the motion of the Earth, we would observe temperatures equal to one part in 10^3. How do these two regions of the Universe, which have never had any interaction up to the time of emission of the radiation, know how to set their temperatures equal to this accuracy? In the big-bang theory we have presented so far, the answer is simply that this was so ordained in the unknown initial 10^{-5} seconds, that is, effectively, in the initial conditions for the Universe. It is difficult to accept that this is really any answer at all.

9. *The Limits of Isotropy*

9.1. Anisotropic Universes

Perhaps the most obvious feature of the Universe accessible to casual inspection is that it is apparently neither isotropic nor homogeneous. Indeed, it required quite a lot of effort (Chapters 1 and 3) to discover that this false characterisation arises because the region of the Universe available to casual observation is insignificantly small. Nevertheless, at some stage we have to face the existence of departures from the exactly isotropic and homogeneous models with which we have so far been concerned. This will enable us to consider such questions as the extent to which the Universe departs from exact symmetry, to characterise the way in which it does so, and to ask whether it could have been more rather than less irregular in the past, or if the irregularities might grow in the future.

The assumption of an exactly Robertson–Walker geometry can be relaxed in two stages. First, we can consider spatially homogeneous but anisotropic models; subsequently, one might attempt the far more difficult investigation of spatially inhomogeneous geometries.

Over sufficiently large scales the Universe appears to depart from spatial homogeneity to only a small degree. If we imagine a region of enhanced density, $\rho + \Delta\rho$, on a length scale l, then the condition

$$\frac{G\Delta M}{l} \sim G\Delta\rho l^2 \ll c^2 \tag{9.1.1}$$

ensures that the enhanced gravitational potential of the extra mass, ΔM, does not induce large, unobserved peculiar velocities in neighbouring matter. If this condition does indeed hold on all length scales, it makes sense to represent the Universe by a model which is exactly homogeneous, and to allow ourselves the possibility of some degree of anisotropy. We can then use observations to set limits on the possible anisotropy. These will turn out to be severe, at least for a large part of the evolution of the Universe. One can then turn to models which are inhomogeneous perturbations of *isotropic* geometries in order to discuss such problems as the evolution of galaxies. If condition (9.1.1) were not valid on some length scale, then the direct or indirect observation of large anisotropic motions in the Universe would yield information on its departure from homogeneity.

There are two aspects to the possible anisotropy of the Universe, relating to the anisotropy of the expansion of matter and the anisotropy of the geometry of space–time, the two being linked, of course, through the field equations of general relativity. Consider first the behaviour of matter.

At each point the particles of matter are following world-lines which are converging or diverging, and possibly rotating about each other. Consider a set of particles which form a sphere at one time. At a later time we shall find the sphere to be distorted (figure 9.1). The sphere may undergo pure expansion with its orientation unchanged, or it may be expanding at different rates in different directions, and so become distorted into an ellipsoid. If it is contracting in some directions in such a way as to maintain a constant volume and orientation, then it is said to be undergoing a purely shearing motion. Alternatively, the sphere may remain unchanged in shape, but its orientation can be changed by rotation about a fixed axis. In the course of time the axes of shear and the axes of rotation may change. The general motion will be some combination of expansion, shear and rotation.

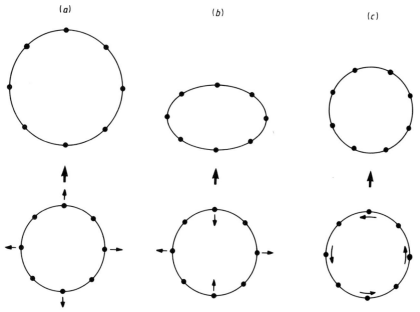

Figure 9.1. The general motion of a sphere of fluid particles, represented here by a circular cross section, is a combination of expansion (*a*), pure shear (*b*), and rotation (*c*).

If the material particles are falling freely we can quantify this description of the motion quite simply. Consider a set of neighbouring fluid particles as seen by an observer attached to one of them. This observer describes the motion he sees in terms of special relativity, using the special coordinates (λ, ξ_i) introduced in §§5.12 and 6.6. In particular, the ξ_i can be treated as standard Cartesian coordi-

197

nates in space. With each neighbouring fluid particle the observer associates a velocity, $\mathbf{u}$. We can write

$$\frac{\partial u_i}{\partial \xi_j} = \frac{1}{2}\left(\frac{\partial u_i}{\partial \xi_j} - \frac{\partial u_j}{\partial \xi_i}\right) + \frac{1}{2}\left(\frac{\partial u_i}{\partial \xi_j} + \frac{\partial u_j}{\partial \xi_i}\right) \tag{9.1.2}$$

$$\equiv \omega_{ij} + \theta_{ij},$$

where the derivatives are evaluated at the observer, $\xi = 0$. By definition, ω_{ij} is the rate of rotation and θ_{ij} the rate of shear at $\xi = 0$. Clearly, $\omega_{ij} = -\omega_{ji}$ and $\theta_{ij} = \theta_{ji}$. We can relate ω_{ij} to the usual fluid dynamical vorticity vector, ω, by

$$\omega_i = \epsilon_{ijk}\,\omega_{jk} \equiv (\nabla \wedge \mathbf{u})_i \,.$$

The quantity θ_{ij} can be further decomposed into a trace and a trace-free part as

$$\theta_{ij} = \tfrac{1}{3}\delta_{ij}\theta + \sigma_{ij} \,,$$

where $\sigma_{ii} = 0$ and $\theta_{ii} = \theta$, summed over i. The expansion θ is just the usual divergence,

$$\theta = \nabla \cdot \mathbf{u} \,.$$

It can be shown that θ, σ_{ij}, ω_{ij} correspond respectively to the expansion, (pure) shear, and rotation of the set of fluid particles discussed above.

Formally, the decomposition of the spatial derivatives of the fluid velocity at a point (9.1.2) exactly parallels the standard treatment of Newtonian fluid dynamics. In the specially chosen coordinate system the equations look exactly the same. The difference between Newtonian fluid dynamics and relativistic cosmology resides in the fact that in the latter a different coordinate system is required at each point. This is technically unmanageable in practice, and so, for the purposes of calculation, methods have been developed using general coordinate systems (e.g. Ellis 1971). The formal similarity with Newtonian theory is then lost, but one should not allow this to obscure the fact that for the suitably chosen observer, expansion, shear and rotation have the same meaning in relativistic cosmology as in classical fluid dynamics.

We have discussed the theory from the point of view of an observer attached to a material particle in free-fall. In fact it can be generalised in two ways. First, the observer need not be in free-fall; pressure forces may prevent the world-lines of matter from being geodesic, so this is an important relaxation. In this case one might ask with respect to what it is that rotation is to be measured. For freely falling particles it is measured with respect to the axes defined by freely falling gyroscopes carried by the observer, or with respect to a frame in which light travels in straight lines. For non-geodesic motion the situation is similar: a non-rotating frame is one in which a light ray is reflected back from a nearby mirror in the same direction as it was emitted. Rotation is measured relative to such a frame, which is the generalisation of a Newtonian non-rotating frame of reference for an accelerated observer.

The second generalisation is to sets of observers not attached to the particles of matter responsible for producing the gravitational fields. For example, even if the particles are not moving geodesically, one might nevertheless wish to consider the expansion, shear and rotation of a set of geodesics. This generalisation is straightforward, since we merely imagine immaterial observers attached to the geodesics and proceed as before.

These considerations apply whether or not the space–time has any special symmetries, since each quantity is defined at a point and can be an arbitrary function of position. We now specialise to the case of a space–time with homogeneous spatial sections, that is, one in which there is some observer not necessarily attached to particles, for whom the set of simultaneous events is homogeneous. Of course, this must be true for a whole set of observers, one through each point of the homogeneous space; so we regain a class of fundamental observers defining a cosmic time. Note that, as usual, these observers cannot verify their status by simple direct observation, since they cannot see simultaneous events. The operational problems involved are non-trivial.

We now have to enquire into the geometry of the homogeneous spatial sections. In the case of isotropic spaces we found there were just three possibilities (labelled $k = 0, \pm 1$ in Chapter 6). Here the situation is a little more complicated. First of all, giving up isotropy is not quite the unique act of renunciation that it sounds to be. A space is isotropic if it looks the same in all directions, but even if it is anisotropic it may still look the same in some continuous subset of directions. If after a certain rotation the space looks to be the same to the fundamental observers, we call that rotation a symmetry transformation. Two rotations, each of which is a symmetry, must give a symmetry transformation when performed together; the inverse of a symmetry transformation must also be a symmetry; and performing no transformation at all is a symmetry. From these conditions it is clear that either space is isotropic or is rotationally symmetric about one fixed direction, or it has no such symmetry. For arbitrary rotations about two axes would generate all possible rotations. If we retain this extra symmetry of rotation about one axis when relinquishing isotropy, we call the resulting space locally rotationally symmetric (LRS). The following discussion applies mainly to the general case in which this extra symmetry is absent.

According to the homogeneity assumption, as we move from point to point in the space we see the same overall picture, so we generate a translational symmetry transformation. Conversely, homogeneity implies that we should be able to get from one point to any other by a symmetry transformation, since all points are equivalent. What then is the difference between different spaces?

To see this consider first a simple example – compare the homogeneity of the two-dimensional sphere and the plane. In a sense this example is too simple because both spaces are LRS (and hence isotropic, of course), but more realistic examples tend to get a bit too complicated. At any point in these spaces there are two independent directions in which one can move to generate a symmetry transformation. We can think of the symmetry as being mapped out by two sets of curves to which these directions are tangent (figure 9.2); in the plane we have chosen translations along two orthogonal directions, and for the sphere we have chosen rotation about two orthogonal directions. The difference between the two spaces on which we wish to focus can be seen immediately in the way the tangent vectors to these curves change in length and direction from point to point. Roughly speaking, we can say that the difference between two homogeneous spaces is characterised by the way we have to turn in order to see the same picture as we move from one point to another in the space.

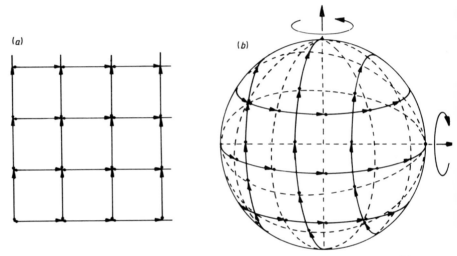

Figure 9.2. Symmetry transformations of the plane (*a*) and the sphere (*b*) generated by vector fields tangent to the integral curves shown. The vector fields change from point to point on the sphere, whereas on the plane they are constant.

We can now describe this mathematically. In three dimensions, spatial homogeneity is described by the existence of three independent sets of curves with tangent vectors ξ_a ($a = 1, 2, 3$). An infinitesimal symmetry transformation takes the point P, coordinates (x_i), to the point P′, coordinates ($x_i + \delta x_i$), where $\delta x = \xi_P \delta t$ for ξ_P some linear combination of the ξ_a at P. The same transformation takes the tip of the infinitesimal vector $\zeta_P \delta s$ at P to $\zeta_{(P+\xi_P\delta s)}\delta t$ at P′; so $\zeta_P \delta s$ becomes $\zeta'_{P'}\delta s = \zeta_P \delta s + \xi_{(P+\zeta_P\delta s)}\delta t - \xi_P \delta t$ in figure 9.3. We can now compare $\zeta'_{P'}$ with $\zeta_{P'}$, the given vector field at P′. The difference between them is called the Lie derivative of ζ along ξ, and is written

$$\mathcal{L}_{\xi}\zeta = \lim_{\delta t \to 0} \frac{\zeta_{P'} - \zeta'_{P'}}{\delta t} = \xi \cdot \nabla \zeta - \zeta \cdot \nabla \xi.$$

Again, roughly speaking, it tells us by how much we must turn a vector carried from P to P′ to make it point in the 'same' direction at P′ as it did at P.

Now, instead of starting with an arbitrary vector at P, we take one of the ξ_a, and instead of going off in an arbitrary direction we move along another of the

200

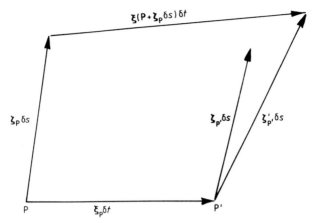

$$\xi(P + \zeta_P \delta s)\, \delta t$$

$$\zeta_P \delta s \qquad \zeta'_{P'} \delta s \qquad \zeta'_{P'} \delta s$$

P $\qquad\qquad$ $\xi_P \delta t$ $\qquad\qquad$ P'

Figure 9.3. The transformation generated by the vector field ξ takes the vector ζ_P at P, to $\zeta'_{P'}$ at P'. The difference between $\zeta_{P'}$ and $\zeta'_{P'}$ gives the Lie derivative (at P) of the vector field ζ along ξ.

ξ_a. The type of space is specified by the non-zero values of the Lie derivatives. We can write

$$\mathcal{L}_{\xi_b} \xi_a = C^c_{ab}\, \xi_c,$$

since the Lie derivative is itself a vector, being the difference between two vectors, and so must be expressible as a linear combination of three independent vectors. It can be shown that the C^c_{ab} are constants, and that they are antisymmetric in the sense that $C^c_{ab} = - C^c_{ba}$. Conventionally we can write

$$C^c_{ab} = \epsilon_{abd} N_{dc} + \delta_{ca} A_b - \delta_{cb} A_a,$$

where N_{dc} is symmetric, this being the natural decomposition of a collection of three antisymmetric matrices. The type of space is specified through the values of N_{dc} and A_a. In fact, we have the freedom to choose new linear combinations of ξ_a as the basis, a freedom which can be exploited to make N_{dc} a diagonal matrix, $(N_{cd}) = \mathrm{diag}\,(N_1, N_2, N_3)$, and $A_a = (A, 0, 0)$, provided $N_1 A = 0$. This leads to the classification scheme set out in table 9.1.

From table 9.1 we see that the basis vectors are on an equal footing in types I, V and IX. We expect to be able to let the shear of the matter tend smoothly to zero in these types and recover the isotropic Robertson-Walker metrics. (In fact, because of the somewhat asymmetric classification scheme, this is also true of types VII$_0$ and VII$_h$.) Types I, V and IX are the simplest generalisations of the $k = 0$, -1, $+1$ Robertson-Walker models, respectively. Just as was the case for isotropic models, some of the simpler anisotropic models for which exact forms of the metric are known are given names. In the next section we consider one of these cases.

Table 9.1. The Bianchi–Behr classification of homogeneous three-dimensional spaces giving the values of the classification parameters (zero, positive or negative) and the type label. The parameter h is defined by $h = A^2/N_2 N_3$.

	Value of parameters in canonical form			
Type of space	A	N_1	N_2	N_3
I	0	0	0	0
II	0	+	0	0
VI_0	0	0	+	−
VII_0	0	0	+	+
VIII	0	−	+	+
IX	0	+	+	+
V	+	0	0	0
IV	+	0	0	+
VI_h	+	0	+	−
VII_h	+	0	+	+

9.2. The Kasner Solution

The simplest example we can consider is the class of Bianchi type I models. It can be shown that the metric for this case can be written as

$$d\tau^2 = dt^2 - X_1^2(t)\, dx_1^2 - X_2^2(t)\, dx_2^2 - X_3^2(t)\, dx_3^2,$$

which is an obvious generalisation of the $k = 0$ Robertson–Walker metric. Clearly, X_1, X_2 and X_3 govern the expansion rates in three orthogonal directions. To determine X_1, X_2 and X_3 as functions of t we need three field equations, one for each of the three orthogonal directions.

We assume that the space–time is filled with a perfect gas with non-zero pressure, with the stipulation that the pressure be a function of time, t, only. Individual particles then have non-zero random motion, but there is no net flow of matter through space, and on average matter follows the geodesics $x_i =$ constant. Introducing $a^3 = X_1 X_2 X_3$, the Einstein equations then give

$$\frac{\ddot{X}_i}{X_i} - \left(\frac{\dot{X}_i}{X_i}\right)^2 + 3\left(\frac{\dot{X}_i}{X_i}\right)\left(\frac{\dot{a}}{a}\right) = \frac{4\pi G}{c^2}\left(\mu - \frac{p}{c^2}\right) \tag{9.2.1}$$

(with no summation over i), and

$$\frac{\dot{X}_1 \dot{X}_2}{X_1 X_2} + \frac{\dot{X}_2 \dot{X}_3}{X_2 X_3} + \frac{\dot{X}_3 \dot{X}_1}{X_3 X_1} = \frac{8\pi G}{c^2}\mu, \tag{9.2.2}$$

which is equivalent to the additional conservation equation.

The expansion in each direction is $\dot{X}_i/X_i$. The mean rate of expansion is

$$\frac{\dot{a}}{a} = \frac{1}{3}\left(\frac{\dot{X}_1}{X_1} + \frac{\dot{X}_2}{X_2} + \frac{\dot{X}_3}{X_3}\right).$$

It is therefore plausible that the shear should turn out to be given by the matrix

$$(\sigma_{ij}) = \begin{pmatrix} \dot{X}_1/X_1 - \dot{a}/a & 0 & 0 \\ 0 & \dot{X}_2/X_2 - \dot{a}/a & 0 \\ 0 & 0 & \dot{X}_3/X_3 - \dot{a}/a \end{pmatrix},$$

that the volume expansion is

$$\theta = 3\dot{a}/a,$$

and that the rotation $\omega_{ij} = 0$. Equations (9.2.1) can be rewritten in the form of evolution equations for σ_i and θ as

$$\dot{\sigma}_i + \theta\sigma_i = 0, \tag{9.2.3}$$

where $\sigma_i = \dot{X}_i/X_i - \dot{a}/a$, and

$$\dot{\theta} = \frac{12\pi G}{c^2}\left(\mu - \frac{p}{c^2}\right) - \theta^2. \tag{9.2.4}$$

There are only three independent equations here since $\sigma_1 + \sigma_2 + \sigma_3 \equiv 0$. Equation (9.2.2) is equivalent to

$$\tfrac{1}{3}\theta^2 = \sigma^2 + \frac{8\pi G}{c^2}\mu, \tag{9.2.5}$$

where $\sigma^2 = \tfrac{1}{2}(\sigma_1^2 + \sigma_2^2 + \sigma_3^2)$, which is the analogue of equation (5.13.6) with the additional term σ^2 representing the energy of shearing motion.

Integrating equation (9.2.3), using $\theta = 3\dot{a}/a$, we obtain

$$\sigma_i = \Sigma_i/a^3, \tag{9.2.6}$$

where the Σ_i are constants, and $\Sigma_1 + \Sigma_2 + \Sigma_3 = 0$.

To proceed further we need an equation of state. For simplicity take the case of dust, $p = 0$, $\mu = \rho$. Then it is easy to show that equations (9.2.3), (9.2.4) and (9.2.5) imply the usual conservation equation

$$(\rho a^3)^{\boldsymbol{\cdot}} = 0,$$

as indeed must follow from the conservation of matter. Therefore $\rho \propto a^{-3}$, whereas from equation (9.2.6) $\sigma^2 \propto a^{-6}$. Consequently, as $a \to 0$ the anisotropy energy dominates the matter term in equation (9.2.5), and the solution approximates to a pure vacuum space–time near what will turn out to be the initial singularity.

If $p = 0$, $\mu = 0$ exactly, we obtain the *Kasner solution*, which is given by

$$X_i = \mathcal{X}_i t^{p_i}, \quad \mathcal{X}_i = \text{constant}, \quad p_1 + p_2 + p_3 = 1 = p_1^2 + p_2^2 + p_3^2.$$

It is easy to check that this satisfies equations (9.2.1) and (9.2.2) by direct substitution. Thus we have $a^3 \propto t^{p_1 + p_2 + p_3}$, so

$$a \propto t^{1/3}, t \to 0. \tag{9.2.7}$$

Consequently, the space–time is singular at $t = 0$ as stated.

Different types of singularity are possible here depending on the relative rates of expansion in different directions. In all cases, the matter density becomes infinite (compare §11.3). If all $X_1, X_2, X_3 \to 0$ as $a \to 0$ we have a 'point' singularity, similar to the isotropic models. If $X_1, X_2 \to 0$ but X_3 remains finite, we have a 'barrel' singularity with the x_3 axis as the axis of the barrel. If $X_1, X_2 \to 0$ and $X_3 \to \infty$ we have a 'cigar' singularity. Finally, if $X_1 \to 0$ and X_2 and X_3 remain finite we have a 'pancake' singularity at $t = 0$.

The behaviour of the shear, $\sigma_i \propto a^{-3}$, means that its importance diminishes relative to matter at late times; in fact, $\sigma/\theta \to 0$ as $t \to \infty$, and $\int_{t_0}^{\infty} \sigma_i \, dt$ is finite for any $t_0 > 0$. The former condition guarantees that the shear becomes dynamically unimportant, and the expansion tends to that of the appropriate Robertson-Walker model. The latter guarantees that the cumulative distortions in the microwave background are small, so that the model 'looks' isotropic at late times.

While the type I anisotropic models form a useful fund of examples, they do not illustrate all the types of behaviour that are possible, even in the absence of rotation and acceleration. We shall note in Chapter 10 that more complicated models do not tend to isotropy at late times, and we shall remark on the more complicated oscillation of the axes of shear near the singularity in type IX models. In Chapter 11 we shall find that a new type of singularity appears if non-geodesic motion of matter is allowed.

9.3. Observational Constraints

Any anisotropy of space-time geometry would in principal appear as an anisotropy in the distribution of galaxies and their recessional velocities. From direct counts of galaxies, Kristian and Sachs (1966) found an upper limit

$$\left(\frac{|\sigma|}{\theta}\right)_0 \lesssim 0.3,$$

while from the redshifts of cluster galaxies, Trendowski found a result of the same order of magnitude (see Peebles 1971).

Observations of an anisotropy in the Hubble expansion of Sc galaxies have been reported by Rubin et al (1976), a result which is sometimes referred to as the Rubin-Ford effect. This could be explained in terms of the anisotropy introduced by the motion of the galaxy relative to the Hubble expansion discussed in §3.4. However, the velocity required is several times larger than that indicated by the microwave results. On the other hand, Stenning and Hartwick (1980) find a similar effect which agrees with the microwave measurements. The source of the discrepancy is not known (see also Jaakkola et al 1976).

There are difficulties attendant upon the observational separation of inhomogeneity and anisotropy, which can be quite subtle, because inhomogeneities need not show up in an obvious way. In fact, the universe can appear to be homogeneous and anisotropic even if it is really locally inhomogeneous but isotropic on a large scale. To see this, consider the following highly unrealistic and oversimplified model which nevertheless illustrates the point.

Suppose we use the apparent magnitudes of supernovae to measure distances to galaxies, calibrating from local observations to an absolute magnitude of $M_V = -18$ at maximum light. Suppose that in two clusters in different directions a sample of supernovae are detected at $m_V = 17$. The obvious deduction is that the clusters are at the same distance from us. Suppose next that redshift measurements give values of $z_1 = 0.033$ and $z_2 = 0.025$ for the two clusters. The natural conclusion is that the expansion of this universe is anisotropic with the Hubble constant varying between 100 and 125 km s^{-1} Mpc^{-1}. However, we have to take into account a spread in the intrinsic luminosities of type I supernovae of about 1 magnitude. An alternative interpretation of the data is that $H = 100$ km s^{-1} Mpc^{-1}, and that one cluster is actually further away by about 25%. The supernovae observations are then explained by an intrinsic spread in brightness. We could be comparing the brightest $M_V = -18$ events in the more distant, larger cluster with the relatively more frequent events, typically $\frac{1}{2}$ magnitude fainter, in the smaller, nearer cluster. In this interpretation the universe is expanding isotropically on average, but is locally inhomogeneous. The discussion here is related to the Malmquist effect, according to which a magnitude-limited sample will tend to select more distant objects from the brighter end of a distribution in absolute magnitude. This effect has to be estimated and allowed for in the interpretation of observations.

A higher degree of precision is available from the microwave data. We can think of the microwave background as providing us with information on conditions in the universe at the surface of last scattering (§8.8). In a low-density universe this is at $z \sim 10^3$, and in the high density case with a reionised intergalactic medium, at $z \sim 7$. The limits deduced for the shearing motion will therefore differ somewhat in the two cases. These also depend on the type of anisotropic model under discussion, since geometrical factors affect the propagation of the radiation. In essence, however, the idea is to compute the temperature pattern expected on the sky for a given velocity distribution of electrons at last scattering, taking into account Doppler effects and gravitational redshifts. In general, one expects a contribution to the anisotropy on a 180° scale, or a component with a 12 h sidereal time variation, which can be distinguished from the 24 h variation due to the motion of the Earth. This is a result of the ellipsoidal symmetry of the shearing motion.

For $\delta T/T \lesssim 10^{-3}$ on a 12 h timescale, Collins and Hawking (1973b) find in the *most* favourable case

$$(\sigma/\theta)_0 \lesssim 6 \times 10^{-8}$$

and, at worst, to order of magnitude,

$$(\sigma/\theta)_0 \lesssim 10^{-3}.$$

In principle, this approach could be developed to yield information on the type of anisotropy. For example, different types of model will give rise to different discrete symmetries, the temperature pattern looking the same in a particular finite set of directions (see MacCallum 1973). In practice this is beyond the realms of present technology, since at present we have only upper limits even for the 12 h variation.

The improvement in the limits on anisotropy obtained from the microwave observations results very largely from the fact that we obtain here information on the Universe at a much earlier time than is available from observations of galaxies. A modest restriction on the shear at early times can represent a severe constraint now, since the shear decays rapidly. If we can extend the expected limits on the allowable shear to even earlier times we might expect even more severe constraints. This can be achieved if we accept that the helium abundance is cosmological.

A large shear at the epoch of helium synthesis would alter the time available for the reactions to occur, since σ^2 contributes to speed up the rate of evolution (equation 9.2.5). If helium must be produced in an abundance of $29 \pm 4\%$ by mass (Peimbert 1973), and the deuterium abundance is not to be excessive, then it turns out that at most a 30% variation in the rate of expansion from the Robertson-Walker rate is permissible. This gives a limit at the time of helium synthesis of

$$(\sigma/\theta)_{\text{He}} \lesssim 0.5$$

(Barrow 1976). Again, what this means for the present shear depends on the model. The type I model with critical density gives

$$(\sigma/\theta)_0 \lesssim 5 \times 10^{-12},$$

while for a low-density, $q_0 = 0.01$, type V model (the analogue of the $k = -1$ Robertson-Walker model) one obtains

$$(\sigma/\theta)_0 \lesssim 2 \times 10^{-11}.$$

In most cases the limits are better than those attainable from the microwave isotropy.

9.4. The Rotating Universe

From the flattening of its poles and from the motion of a Foucault pendulum we know that the Earth is rotating relative to a frame of reference in which Newton's laws hold, i.e. relative to a dynamical inertial frame. To choose the Earth as a rest frame, as we do in the pursuit of our daily lives, is therefore dynamically inaccurate to about one revolution in 24 h. We can be rather more accurate than this, since we know that the solar system obeys Newtonian dynamics to a good approximation. To choose the outer planet, Pluto, as a state of rest is therefore dynamically inaccurate to about one revolution in one Plutonian year (about 250 terrestrial years). The flattening of the Galaxy shows it to be rotating, so its outer stars are fixed to about one revolution in one Galactic year (about 2×10^8 terrestrial years). The development of this argument is clear. The important point is that as we go further out there is a steady reduction in the relative angular velocity between the materially defined standard of rest, or the kinematic inertial frame to give it its full title, and the dynamical inertial frame. The argument comes to a halt with our velocity relative to the Local Supercluster, which we would attribute to an angular rotation of at most one revolution in 10^{10} years.

When Galileo needed an inertial frame for the theory of the tides; when, later, Newton needed an inertial frame for the theory of planetary orbits, it was with remarkable serendipity that they should have chosen the 'fixed' stars. For classical dynamics contains nothing that requires even the existence of any stars, let alone their coincidence with the dynamical inertial frame. To a large measure it is this coincidence that gives an attractiveness to Mach's Principle, according to which the local inertial frame is supposed to be actually determined by the matter in the Universe, and hence by the stars. However, sceptics might observe that one revolution per 10^{10} years is just about one revolution in the lifetime of the Universe. This is equivalent to a velocity of light at the edge of the Universe, a distance c/H_0 away. One might argue that the existence of galaxies situated, to order of magnitude, at a distance c/H_0 which do not show large transverse Doppler shifts means that the Universe cannot be rotating faster than at about this rate anyway. In other words, the sceptical response might be that one revolution per lifetime is what one might expect a well disciplined universe to do.

The stronger limits that can be obtained from observations of the microwave background are therefore of considerable interest. First of all, the homogeneity of the Universe implies that distant galaxies, on average, cannot be moving through the microwave background faster than we are. Thus the upper limit for the rotational velocity of matter a distance c/H_0 away comes down to about 600 km s^{-1}, instead of c, and the angular rotation rate to about one revolution per 10^{13} years (10^{-6} per century). Again, however, one can improve on this since non-random electron velocities at the time of last scattering would distort the temperature distribution on the sky in a calculable way. The improvement arises since this constrains the velocities at high redshifts, and these decay as $v \propto (1+z)^{-1}$ (§8.2). Thus Hawking (1969) found a further improvement on the limit of about one and a half orders of magnitude at the very least, or an upper limit of one revolution per 3×10^{14} years.

It might be thought that this low rate could be accounted for in terms of the conservation of angular momentum, which implies $\omega \propto a^{-2}$ as the Universe expands. In fact this is not the case; it can be shown that $\omega \ll H$ at all times for which a spatially homogeneous model is a good approximation (Collins and Hawking 1973b). Observations of the microwave background therefore show that the rotation of the Universe is exceedingly slow in this well defined sense (and is compatible with zero rotation). Thus in the problem of accounting for the high degree of isotropy of the Universe, to be discussed in the next chapter, we include the problem of its slow rotation.

9.5. The Inhomogeneous Universe

The exact solution of the Einstein field equations for the geometry of a space–time containing a general inhomogeneous distribution of matter is not possible because of the nonlinear complexity of the equations. Exact models of inhomogeneous universes are known, but only for unrealistically symmetric matter distributions (e.g. Bondi 1947, see MacCallum 1979). One might imagine that the treatment of small departures from homogeneity, where nonlinearities can be ignored to a first approximation, would not present too much difficulty. Consider then the metric

$$ds^2 = \left(1 + \frac{\epsilon}{1+\tau^2}\right)^2 d\tau^2 - \left(1 + \frac{\epsilon}{1+\xi^2}\right)^2 d\xi^2 - \left(1 + \frac{\epsilon}{1+\eta^2}\right)^2 d\eta^2 - \left(1 + \frac{\epsilon}{1+\zeta^2}\right)^2 d\zeta^2,$$

where ϵ is a small parameter. This looks just like a small departure from Minkowski space–time, due perhaps to some dilute distribution of matter. In fact it is not a perturbation at all, but Minkowski space–time written in an unhelpful coordinate system. This can be checked explicitly without too much difficulty here, but in general the problem of small perturbations is complicated in general relativity by the need to distinguish physical effects from changes in coordinates. A second, partly related difficulty involves the separation of gravitational radiation resulting from the interaction of lumps of matter which may be present in the perturbed space–time, from source-free gravitational radiation which in most cases one would like to exclude.

A further problem involves the relation between a solution of the linearised equations for a perturbation and an exact solution of the full equations (D'Eath 1976). The problem is that there may be no relation at all: the linearised solution may approximate no exact solution whatever. For example, one might think that the simplest way to approach slowly rotating models is to consider a small rotational perturbation with no shear. However, this would correspond to no exact solution at all, since in the full theory vanishing shear implies vanishing rotation if the space is expanding.

Nevertheless, partial discussions of small inhomogeneities in otherwise isotropic universes have been given, particularly with reference to the effect on the microwave background. We can distinguish here between large-scale inhomogeneity, for which general relativity is necessary, from small-scale local clumpiness, for which it is not. Consider first the large-scale effects.

Suppose we live in a density perturbation of order $\delta\rho$ on a length scale l, and suppose that radiation is not scattered on its way to us through this material. The extra matter will nevertheless have a gravitational effect which turns out to be of order (Sachs and Wolfe 1967):

$$\frac{\delta T}{T} \sim \frac{\delta\rho}{\rho} \cdot \frac{l^2}{(c/H_0)^2}$$

for $q_0 = \frac{1}{2}$. This expression is just the ratio $\delta\phi/\phi$ of the gravitational potential due to the lump at its edge, to the gravitational potential of the visible Universe. Observations give $\delta T/T \lesssim 10^{-3}$, so for $l \sim \frac{2}{5} c/H_0$, say, which corresponds to a red-shift of $z \sim 0.4$ and length scale of about 1000 Mpc, we have $\delta\rho/\rho \lesssim 5 \times 10^{-3}$. This particularly low value depends on the location of the perturbation and the assumption that $q_0 = \frac{1}{2}$. Nevertheless, the method may be contrasted with Hubble's counts of galaxies which gave $\delta\rho/\rho \lesssim 2$ on this scale (§1.2).

An alternative possibility is large-scale clustering of the matter through which the background radiation must pass on its way to us. Here one is thinking of, say, clusters of galaxies on scales of about 1000 Mpc again, but with density contrast $\delta\rho/\rho \sim 2$, so a perturbation analysis does not apply. A neat way of estimating the effect in this case is the so-called *Swiss cheese* model, which is a very special exact inhomogeneous cosmological model.

Within a spherical region of a Robertson–Walker space–time we can rearrange the material in any spherically symmetric manner without affecting the space–time outside that region, provided only that we do not alter the total mass. This is an analogue of the Newtonian result that any spherical distribution of matter produces an external gravitational field equal to that of a point mass at its centre. In particular, one can imagine a region which has been pushed inwards and has therefore separated off with a higher density from the expanding substratum. Furthermore, since they do not communicate their presence to the exterior space–time, the model may contain as many of these reconstituted holes as one likes. This is the Swiss cheese cosmological model.

Because each hole exists in an expanding environment, the passage of a light ray through the hole is not symmetrical, and this gives rise to a shift in frequency. Such holes should therefore make themselves visible by the imposition of a characteristic temperature profile on the microwave background (Rees and Sciama 1968, Dyer 1976). A mass of $10^{19} M_\odot$ in a region of diameter 750 Mpc with a density contrast $\delta\rho/\rho \sim 2$ and situated at $z \sim 1.5$ would subtend an angle of $20°$ across which the temperature would vary by a few tenths of a per cent. This would be on the verge of detectability – if one knew where to look.

On a much smaller scale, the ionised gas in rich clusters of galaxies, assuming this is the correct interpretation of the x-ray emission, may be expected to produce a perturbation in the microwave temperature. This is a result of the energy exchange by Compton scattering between the hot electrons in the gas and the microwave photons passing through it. The relevant parameter is

$$ y \sim \left(\frac{kT_e}{m_e c^2}\right) \tau_{es} = \frac{kT_e}{m_e c^2} \sigma_T n_e l, $$

where $l \sim 1$ Mpc is the diameter of the x-ray emitting core of the cluster, and l/cy is the timescale for the radiation temperature to change significantly, as in §8.6. To account for the x-ray emission we need $n_e \sim 10^{-3}$ and $T \sim 10^8$ K, and this gives $y \sim 10^{-4}$. The actual change in the microwave spectrum is given by the appropriate kinetic equation including a collision term representing the Compton energy exchange (in this case often called the Kompaneets equation). The result is that the radiation temperature on the Rayleigh–Jeans part of the spectrum is reduced by a factor e^{-2y}, so $\delta T/T \sim 10^{-4}$ (Zeldovich and Sunyaev 1969). Claims have been made for the positive detection of an effect of this order of magnitude in several rich clusters (Gull and Northover 1976), but the subject is still a matter of debate (see e.g. Perrenod and Lada 1979, Birkinshaw *et al* 1978).

10. Why is the Universe Isotropic?

10.1. Beyond Isotropy

The laws of physics are laws in two senses. They are laws in the sense that they provide regulatory principles by which we might hope to organise the chaos of existence. But the laws are in general expressed as differential equations and these cannot tell us what will happen or has happened, but only what cannot happen, because we have to supply initial conditions. They are therefore laws also in the legal sense of providing constraints on our freedom of action. There is little doubt that we shall eventually be able to use our knowledge of physical laws to understand and to explain the way in which the Universe evolves, the interaction of its constituents, the assembly of complexity and its dissolution. We have already seen the sort of form that such an understanding of all but the first 10^{-5} seconds of our physical history might take, even if we do not yet possess all the details. By the acquisition of a deeper knowledge of elementary particles and their interactions, we might hope to understand more of that initial 10^{-5} seconds. But just because of the dual nature of the laws, the application of the theories of laboratory physics to the Universe as a whole unveils the possibility of other universes, and the possibility that our Universe might have been quite different.

Even if a response to this problem in its full generality is possible at all, it is not yet fruitful to attempt it, but we can particularise the problem in the present context as an appeal for an explanation of the high degree of isotropy of our present Universe. There are several possible responses which are explored in more detail in the following sections.

A particular symmetry is a particularly appropriate assumption if the system under discussion tends to acquire that symmetry from any reasonable starting conditions anyway. Thus one might try to start the Universe with arbitrary anisotropy and hope that it would tend to isotropise. To this end one might attempt to use only conventional physics (some known form of viscous damping, for example), or one might try to introduce processes operating before 10^{-5} seconds, about which a certain amount of speculation – or wishful thinking – is required (§10.3).

A necessary condition for a symmetry assumption to be appropriate is that it is not destroyed by the unlimited growth of small perturbations. Of course, in our real world, all we know is that isotropy has not *yet* been destroyed, but it is not unreasonable to assume we are not being misled in what would be a particularly malicious manner. If in general it were to be found that perturbations do not grow, and hence that the general model tends to isotropise, we might answer the question as to why the Universe now appears isotropic by noting that it would have to be a very special, hence unlikely, Universe that did not. On the other hand, we should not be surprised to find that we live in a special (isotropic) Universe if the physical basis for the existence of life of sufficient intelligence to pose these questions (or of sufficient stupidity to fail to answer them), were to be that the Universe must be somewhat special (§10.4).

An alternative approach to the problem is to suppose that the incompleteness of our physical laws allows them to admit too much. If only we knew the whole of physics, one might argue, then we would see that only symmetric initial conditions were possible for there to exist any Universe at all: the phoenix of a Universe arising out of its unity (Sciama 1959). This is just the sort of approach Einstein had in mind in his discussion of Mach's Principle (§6.1). Einstein had hoped that the equations of general relativity would admit no solution at all, hence no geometry to the Universe and no Universe, in the absence of matter. For this is a necessary condition that geometry and inertia arise as the consequence of matter, i.e. a necessary condition for Mach's Principle to be satisfied. For reasons to be given in §10.2 one might hope also that Mach's Principle would rule out anisotropic models. In this way we should arrive at the rather satisfactory conclusion that the Universe is isotropic because there are no other possibilities.

10.2. Mach's Principle and Isotropy

A not unreasonable initial reaction to the development of relativistic cosmology is that, far from limiting the structure of the world, general relativity provides a plethora of hitherto unimagined possibilities. One can think of these as arising from the imposition of different initial conditions for the differential equations which are provided by general relativity as the field equations for the geometry of space–time. The simplest way to limit the possibilities, therefore, is to find some natural limitation on the initial conditions. The first suggestion as to how this might be done came from Einstein himself. It is related to the possibility of implementing Mach's ideas as natural restrictions on the allowable initial conditions. The idea is that Mach's Principle becomes a selection rule allowing, by definition, only 'physically admissible' solutions of the differential equations. After all, we often discard mathematically possible solutions on the grounds of manifest physical implausibility.

One example where Mach's Principle is violated is in a universe containing, say, a single star. Far away from the star, where its influence must vanish, particles behave in a perfectly reasonable way (according to the laws of special relativity, in fact) and the geometry of space–time is quite normal. Therefore, in this case, the geometry and the motion of test particles are not determined by other matter. Einstein argued that such a situation could not arise if the Universe were such that it is impossible to get indefinitely far away from matter. This would be the case in a spatially closed universe. He therefore proposed spatial closure as a necessary initial condition for physically acceptable cosmological models.

This has led some people to think that the Universe must be spatially closed. Unfortunately, as we have seen in Chapters 4 and 9, this belief is in danger of conflicting with observation. If that conflict were established, then this form of Mach's Principle would be of little help in restricting the range of possible models. In addition, there exist spatially closed models without any gravitating matter in which, therefore, the behaviour of test particles cannot be determined by matter, and Mach's Principle is not satisfied. Therefore, even if closure is a necessary requirement of Mach's Principle, it cannot be sufficient. In pursuing this approach one is therefore forced to look for alternative means of restricting the initial conditions.

There are some grounds for hoping that a successful development of these ideas would help to explain the isotropy of the Universe. We know that Mach's Principle forbids rotating-universe models (§9.4), and therefore to this extent forbids a certain anisotropy. One can also argue that a universe in which the matter were shearing but not rotating should also be forbidden in a correct implementation of Mach's Principle (Bondi 1960). For the shearing motion would mimic a rotation of the Universe when viewed from an Earth stationary with respect to an inertial frame, and therefore, according to Mach, supposedly stationary with respect to the Universe! (figure 10.1). It therefore follows that Mach's Principle should rule out the introduction of shear and rotation, and hence of anisotropy, into a spatially homogeneous model. A correct implementation of Mach's ideas in general relativity would presumably account for any deviation from exact isotropy in the Universe in terms of small deviations from exact homogeneity.

How then can we complete general relativity in a systematic way to incorporate Mach's Principle? One way of achieving this can be seen by comparison with the manner in which boundary conditions are introduced into Newtonian gravitational theory. Here the field equations are usually stated in the form of Poisson's equation,

$$\nabla^2 \phi = 4\pi G \rho,$$

and the usual boundary condition, $\phi \rightarrow 0$ at infinity, specified separately. But if,

Figure 10.1. A shearing motion of the stars can mimic a rotation to a certain extent.

instead, we write the field equations in the integral form

$$\phi(\mathbf{r}) = 4\pi G \int \frac{\rho(\mathbf{r}')\, d^3\mathbf{r}'}{|\mathbf{r}-\mathbf{r}'|},$$

the boundary conditions have been incorporated automatically, and the gravitational potential is manifestly determined by the matter distribution.

In general relativity, the role of ϕ is played by the metric coefficients, $g_{\mu\nu}$, and the gravitational effect of matter is determined by its stress–energy density, $T_{\mu\nu}$, not just its mass density. Since we are dealing with a relativistic theory we expect the geometry at an event to depend on the matter distribution in the past light cone of the event, and hence that the volume integral be replaced by an integral over space and time. Furthermore, since the geometry of space–time is not Euclidean, we expect the function $|\mathbf{r}-\mathbf{r}'|$ to be replaced by an expression which itself depends on the geometry, and hence on $g_{\mu\nu}$. The best we can hope for, therefore, is to replace the differential equations of Einstein's theory by integral equations in which the unknown metric appears, both explicitly and in the integral. This turns out to be possible and the result is equivalent to general relativity, except that in this new form the initial conditions, or boundary conditions, are automatically incorporated into the theory, and the geometry is manifestly dependent only on the matter distribution (Sciama *et al* 1969, Gilman 1970). There are some technical complications, but it turns out that this integral form can be achieved in such a way that each element of matter in the Universe contributes an effect which propagates to us self-consistently through just that space–time geometry that all the matter adds together to produce. This is a necessary requirement, because the total effect of the matter is not the sum of the effects that each piece would produce acting alone. The result would appear to be a theory in which the removal of the Universe around us would have an effect on the local behaviour of matter. If in this theory the matter is removed, one is left not with an empty universe, but with no universe at all, for there is now no consistent empty space–time solution.

Unfortunately, this formulation of Mach's Principle cannot be quite correct. The error, paradoxically, arises from the assumption that only the matter in the past of an event can influence the geometry there. This assumption breaks down in universes which contain horizons. To see this, suppose an extra star to be added to the Universe outside our horizon. The consequences can be calculated exactly according to general relativity, and they confirm the approximate picture we derive from a simple Newtonian argument. To a first approximation, the star exerts its $1/r^2$ gravitational force, and pulls the Universe towards it. This pull can be felt even before the star can be seen. The star cannot signal to us, so we cannot know what type of star it is, or indeed anything about the mass distribution beyond its spherical symmetry. Therefore, the requirement of 'causality' is not violated. One might be tempted to ask what would happen if the mass were to disappear; whether the instantaneous release from its attraction would be non-causal. Indeed it would, but it cannot happen, because mass must be conserved. The conservation of mass

214

appears here as an aspect of causality! Note that there is no conflict with our discussion of cosmological principles. If the theory were indubitably true we could deduce the existence of any matter beyond our horizon; but the theory could only be confirmed as true if we were to know this matter to exist!

As a consequence the correct formulation of Mach's Principle should allow for the influence of matter beyond the horizon. So far this has been achieved only at the expense of a lot of technical machinery (Raine 1975). The resulting theory appears to be successful in ruling out spatially homogeneous anisotropic models. It may therefore provide a step towards the explanation of the isotropy of the Universe. It does not as yet offer any explanation as to why the Universe should be approximately homogeneous; nor, one cannot but feel, is the theory sufficiently beautiful to be completely true.

Whatever the ultimate mathematical success of this approach, there are still profound difficulties of a philosophical nature. For we have modified general relativity in a way that does not change laboratory physics, but limits only the cosmological possibilities. We have therefore no way of testing this modification. Provided that our laws allow the Universe in which we do live to exist, we have no way of knowing whether they correctly forbid the possibility of those in which we do not. It appears that in this way we explain everything at the cost of explaining nothing.

10.3. Chaotic Cosmology

The simplest resolution of the problem posed by the isotropy of the Universe would be that it arises as a consequence of the normal operation of the laws of physics on a state of arbitrary chaos. In this respect the tendency to isotropise might be compared with the tendency of physical systems to evolve towards equilibrium. In order for this to happen the Universe has to ensure that it has had sufficient time to dissipate an arbitrarily large amount of anisotropy, and to provide itself with a mechanism for so doing.

This programme has so far involved two different but related aspects. The first aspect concerns the elimination of horizons (Misner 1969). Consider a particularly simple example of a type I metric,

$$d\tau^2 = dt^2 - t^2 dx_1^2 - dx_2^2 - dx_3^2, \qquad (10.3.1)$$

which is really a part of Minkowski space–time in disguise, but nevertheless serves to illustrate the point to be made. The horizon is now different in different directions. We compute the coordinate values of x_1, x_2, x_3 at the horizon of an observer at the spatial origin at time t, by tracing back an incoming light pulse to the singularity at $t = 0$. For this we need the equations of light rays. Either from the general theory, or by transforming to normal coordinates, these are

$$t^2 \frac{dx_1}{d\lambda} = p_1, \qquad \frac{dx_2}{d\lambda} = p_2, \qquad \frac{dx_3}{d\lambda} = p_3,$$

where p_1, p_2, p_3 are constants of integration. The parameter λ is determined by the condition that a light pulse follow a null geodesic:

$$0 = \left(\frac{dt}{d\lambda}\right)^2 - \frac{p_1^2}{t^2} - p_2^2 - p_3^2,$$

from (10.3.1). These equations integrate to

$$x_1 = \int_\epsilon^{t_0} \frac{p_1\, dt}{t^2(t^{-2}p_1^2 + p_2^2 + p_3^2)^{1/2}}, \qquad x_2 = \int_\epsilon^{t_0} \frac{p_2\, dt}{(t^{-2}p_1^2 + p_2^2 + p_3^2)^{1/2}},$$

$$x_3 = \int_\epsilon^{t_0} \frac{p_3\, dt}{(t^{-2}p_1^2 + p_2^2 + p_3^2)^{1/2}},$$

for a ray received at the origin at time $t = t_0$, and emitted from (x_1, x_2, x_3) at $t = \epsilon$. As $\epsilon \to 0$, x_2 and x_3 converge, but x_1 diverges. In the x_1 direction the observer can see all the galaxies and there is no horizon.

In this model the horizon is removed in one direction, along which the scale factor goes to zero as $t \to 0$. In the orthogonal directions the scale factor remains finite, so we have a 'pancake' singularity (§9.2). To remove horizons completely we need pancake behaviour in each of three orthogonal distinguished directions in turn, and the model must remain in each of these three configurations for a sufficient time.

This behaviour does *not* occur in any type I models, so it is certainly not possible to remove horizons completely in at least these particular examples. At best we can ask whether horizons can be removed in most models, with limited exceptions. Type I models are rather special examples, as a glance at table 9.1 shows. One can think of this in the same way that one conceives the $k = 0$ Robertson–Walker dust model as rather special, since there is only one such model, whereas for $k = \pm 1$ there are an infinite number of different possibilities depending on the value of q_0. Type IX models constitute a more general class. If most type IX models do not remove horizons then it is certainly not the case that the removal of horizons is a general property with limited exceptions. The evolution of the ratios of the shear in different directions can be followed in the type IX models starting from arbitrary initial values to see whether the ratios look like the pancake configuration often enough. It turns out that in most cases they do not, and in general horizons cannot be removed. It would seem that the communication of physical conditions over the whole Universe is not in general possible, and cannot be invoked to account for the observed isotropy. One might argue, however, that the restriction to anisotropic homogeneous models means that the chosen set of initial states is not chaotic enough, and that the question as to whether inhomogeneous models might in general isotropise remains open.

The second aspect of the idea of a chaotic cosmology focuses attention on a mechanism for the dissipation of anisotropy in a spatially homogeneous universe. Dissipation may occur if the matter content does not behave as a perfect gas or if it behaves as a perfect gas not in equilibrium. As long as collisions occur on a timescale much less than the expansion timescale, the matter will remain in equilibrium. Non-equilibrium behaviour occurs as one constituent of the Universe decouples from the rest. The recombination epoch is rather too recent to admit the dissipation of arbitrary amounts of shear, so the next candidate, going back in time, is the decoupling of neutrinos (§8.4). Detailed calculation shows that the viscosity of the neutrinos during this stage cannot dissipate shear anisotropy fast enough to reduce it to zero (Stewart 1968). This happens despite the fact that the rate of dissipation depends on the shear of the gas, because so too does the rate of expansion through this phase (equation 9.2.5).

If σ_ν is a mean cross section for the various neutrino scattering processes, and n_e is the electron density, the neutrino collision timescale is $t_\nu \sim (\sigma_\nu n_e v_e)^{-1} \propto (\sigma_\nu n)^{-1}$, since the electron velocity v_e is approximately c. The Kasner model provides an illustrative example of a highly anisotropic universe in which matter is initially unimportant. In this model we found $n \propto t$ as $t \to 0$ (equation 9.2.7), so $t_\nu/t_{exp} \propto \sigma^{-1}$. Straightforward extrapolation to high energies of the known low-temperature behaviour, $\sigma_\nu \propto T_\nu$, implies $t_\nu \ll t_{exp}$. In this case the neutrinos must be in equilibrium at arbitrarily early times, and the matter content of the early universe behaves as a perfect fluid. If, however, the cross section for neutrino collisions decreases at very high energies, in accordance with certain theoretical predictions, then the neutrino gas becomes collisionless at sufficiently early times, and will not come into equilibrium (Collins and Stewart 1971). An anisotropic early universe cannot then be treated as a perfect gas.

To see what can happen, consider the Bianchi type I models. Typical behaviour here is an overall expansion superimposed on a shearing motion consisting of expansion along two orthogonal directions, and collapse along a third. The neutrinos moving along this third axis become relatively more energetic to the extent that the timescale for neutrino–antineutrino annihilation, which is governed by a cross section roughly proportional to T_ν^2, becomes less than the expansion timescale. Neutrino–antineutrino pairs therefore annihilate to yield an isotropic distribution of electron–positron pairs:

$$\bar{\nu}_e + \nu_e \to e^+ + e^- .$$

The collapse along the third axis is halted at this stage, and the universe expands out along this axis again. Later the process can be repeated, but the annihilation of neutrinos has removed all the anisotropy energy after this one 'bounce', and the model becomes essentially isotropic.

This works in type I models, but in other anisotropic models the anisotropy appears to be regenerated, whereas the conditions to remove it cannot be because of the overall expansion and cooling. The conclusion seems to be that an arbitrarily large initial anisotropy cannot in general be dissipated by this mechanism.

However, enthusiasts for this approach will not want to give up at the first attempt to find an isotropisation time in a universe as ancient as 10^{-5} seconds. Rather than considering the most recent time at which the scheme might work, one could look for the earliest. This might be taken as the time at which gravity is important in the quantum description of matter. One might guess this to be the time, t_h, at which a quantum mechanical particle just fits into a horizon. Estimating the size of a particle by its Compton wavelength, λ_C, this means, very roughly, $ct_h \sim \lambda_C = \hbar/mc$. For protons we get $t_h \sim 10^{-24}$ s. Of course, an alternative proposal would be the epoch when the quantum nature of the gravitational field is important. We can guess this time by constructing (uniquely) a time out of the constants G, h and c, representing the mixture of gravity, quantum theory and relativity. This is the *Planck time*, $(Gh/c^5)^{1/2} \approx 10^{-43}$ s. However, at this stage we do not even know whether we are posing the question in sensible terms, since we do not have a quantum theory of gravity. Rather, the discussion shows that there is an epoch during which quantum gravity can be ignored but the effect of space–time curvature on quantum theory cannot.

We can imagine the quantum vacuum to consist of particle–antiparticle pairs in a constant state of creation and destruction. This is sanctioned by the Uncertainty Principle, which allows the spontaneous appearance of an energy ΔE for a time $\Delta t = \hbar/\Delta E$. It is also a reasonable picture of the mathematical description. Such short-lived particles are called 'virtual'. Suppose we now introduce into the vacuum an external electrostatic field; the theory of quantum electrodynamics allows us to work out what happens. We are interested in the case that the quantum nature of the external field can be neglected and it can be treated classically. It is found that, for sufficiently strong fields, real electron–positron pairs, for example, should be produced. This cannot be checked by experiment, since the rate of particle creation is immeasurably small for any laboratory fields, but it is a rigorous consequence of a theory that is in other respects remarkably accurate in its predictions. We can understand what is happening if we imagine the effect of the electric field on virtual pairs of charged particles. A positron and electron are pulled in opposite directions by the field. If they are separated by much more than a Compton wavelength, $\hbar/m_e c$, in a time $\Delta t = \hbar/\Delta E = \hbar/m_e c^2$, they cannot get together again and annihilate, and so they appear as real particles. The energy comes from the batteries maintaining the field, or from a decay in the field, if it is not maintained. Note then that an anisotropic field is being dissipated in the production of particle–antiparticle pairs.

There is little doubt that a similar process will occur when the quantum mechanical vacuum is disturbed by a non-quantised gravitational field. Of course, in this case, the field affects both charged and uncharged particles alike. It pulls each of a virtual pair of particles in the same direction, but along geodesics which diverge. If the paths diverge sufficiently rapidly the above argument applies. We can estimate the time, t_q, at which this will occur in the expanding universe from the geodesic deviation equation. For the relative acceleration, $\ddot{\xi}$, of a pair of geodesics separated by ξ, equation (6.6.1) gives

$$\frac{\ddot{\xi}}{\xi} \sim \frac{\ddot{a}}{a} + \left(\frac{\dot{a}}{a}\right)^2 \frac{\dot{\xi}^2}{c^2}$$

to order of magnitude. Let $\Delta\xi$ be the separation achieved in a time Δt. The left-hand side has order of magnitude $(\Delta t)^{-2}$, while the terms on the right are of order t_{exp}^{-2} and $t_{exp}^{-2}(\Delta\xi/\Delta t)^2 1/c^2$. With the usual estimate $t_{exp} \sim t_q$ at time t_q, these terms balance if $\Delta t \sim t_q$ and

$$\Delta\xi \sim ct_{exp} \sim ct_q .$$

The condition $\Delta\xi \sim \hbar/mc$, that the pairs diverge sufficiently to become real, becomes exactly the condition for the virtual particle to fit inside its horizon, i.e. $t_q \sim t_h$. We conclude that at this stage copious particle production is expected to occur.

One might hope that this would provide a mechanism for the dissipation of anisotropy as in the electrodynamic case. To confirm this a proper theory is required, since it is not inconceivable that an anisotropic generation of particle pairs could produce stresses that would regenerate an anisotropic expansion. In addition we need the dissipation to be rapid enough to work for arbitrary initial conditions. Unfortunately there is no agreement on a correct general relativistic quantum field theory in the cosmological context, and no hope of testing putative theories by experiment. Claims have been made for the dissipation of anisotropy by particle creation, but these appear to depend on so far unjustifiable assumptions concerning the quantum state of the system prior to the particle production epoch.

A word of caution is in order at this point. The creation processes described here involve particle–antiparticle pairs, and are a consequence of 'conventional' physics. This is quite different from the creation process postulated in the steady state model, in which particles are produced without their antiparticles, thereby violating a fundamental conservation law of the standard theory. One might ask whether the quantum mechanical process could be called upon instead to provide the continuous creation required by the steady state theory. Essentially because of the rather special nature of the space–time of the steady state theory (Chapter 11), it appears that pair creation in the required amount is possible (Candelas and Raine 1976). However, the subsequent inevitable annihilation of particle–anti-particle pairs would produce a gamma ray flux at the Earth many orders of magnitude greater than the observed background, so this variant of the steady state theory is also ruled out.

A rather different approach can be taken to the development of an isotropic universe from an irregular initial state. Here one concentrates on the fact that the dissipation of structure produces entropy for which the natural repository is the microwave background. Instead of requiring the dissipation of arbitrary anisotropy to yield our smooth world, we imagine that the amount of irregularity to be dissipated early on is just that required to provide the entropy of the background radiation. Of course, if all that is demanded is the generation of one number $(s \sim 10^8 k/q_0)$ from one other number (the degree of irregularity in some sense), success can come in many ways. To determine the likely scenarios the outcome must be related to other features of the universe.

In one of the earliest examples of this approach, Zeldovich (1972) noted that the range of structure on the scales of galaxies to clusters appears to correspond to a density contrast $\delta\rho/\rho \sim 10^{-4}$ at the time at which the scale just fits inside the horizon. This could arise from a natural range of irregularities arising at the end of the quantum gravity era at $t \sim 10^{-43}$ s, in which a region of mass m was involved in a density contrast of $\delta\rho/\rho \sim 10^{-4} (m/M_P)^{-2/3}$, where $M_P = (ch/G)^{1/2} \approx 10^{-8}$ kg is the *Planck mass* (the unique mass constructed from G, h and c). The dissipation of these irregularities on scales less than Galactic size appears to give $s \sim 10^8 k$. Schemes for the production of at least part of the microwave background by dissipative processes are now an active area of research. The possible role of grand unified theories is considered briefly in §12.2.

Turning the argument round again, it can be seen that there is a serious difficulty in the original concept of chaotic cosmology as it concerns the dissipation of arbitrarily large anisotropies. For if the anisotropy energy has to appear as already detected entropy, and one would need considerable cunning to violate this proviso, there is a serious constraint on how much anisotropy can be dissipated. So serious is this constraint that it can be used to provide a stringent limit on the present anisotropy of the Universe! (Barrow and Matzner 1977).

10.4. The Anthropic Principle

The Universe seems to derive its interest for us from a large number of apparent coincidences. Small changes in nuclear properties could remove the variety of elements leaving only universal helium or iron; small changes in atomic properties could prevent the existence of stars like the Sun, with its life-supporting potential; or leave us without the four DNA bases with their particularly subtle spatial structures.

Cosmology too is not without its coincidences. The best known are the approximate numerical equality of certain apparently unrelated large numbers. The fact that these numbers are so large makes their agreement even to within several orders of magnitude highly significant. For example, the ratio of the electric to gravitational force between two electrons is

$$\frac{e^2}{Gm_e^2} \sim 10^{40}. \tag{10.4.1}$$

The size of the visible Universe in units of the classical radius of the electron is

$$\frac{c/H_0}{e^2/m_e c^2} \sim 10^{40}. \tag{10.4.2}$$

The square root of the number of particles in the visible Universe is

$$\left[\frac{4\pi n}{3}\left(\frac{c}{H_0}\right)^3\right]^{1/2} \sim 10^{40}. \tag{10.4.3}$$

Now (10.4.1) is, we suppose, a fixed number independent of time, whereas (10.4.2), (10.4.3) depend on time in different ways. Their approximate equality must therefore be an accident of the time of writing. Cosmologists of 10^{10} years hence, or 10^{10} years ago, will not, or would not be puzzled by these mysterious equalities. This, of course, gives us a clue to their explanation: for 10^{10} years ago there were no cosmologists, here or anywhere else, because there were no stars, and 10^{10} years or so hence there will be no cosmologists because no stars will have habitable planets (or, at least, naturally occurring ones). We should therefore try to account for these coincidences as a precondition of our existence. This can be done in the following way.

The mass of a star is, within two orders of magnitude or so, the maximum not subject to disruption by radiation pressure, which can be shown to be roughly

$$50 \left(\frac{e^2}{hc} \right)^{-3/2} \left(\frac{e^2}{Gm_e} \right)^{3/2} m_p .$$

The lifetime of a star of this mass, which is the shortest stellar lifetime, is

$$(t_*)_{min} \sim 10 \left(\frac{e^2}{m_e c^2} \right) \left(\frac{e^2}{Gm_e^2} \right) . \tag{10.4.4}$$

We might imagine that for evolution of intelligent life we require $t \gtrsim (t_*)_{min}$ for the generation of heavy elements by Population II stars, and $t \lesssim (t_*)_{max}$ before the death of a galaxy. Since, to sufficient accuracy, $t_* \propto M^{-2}$, we have $10^4 (t_*)_{min} \gtrsim t \gtrsim (t_*)_{min}$. Setting $t = c/H_0$, and using (10.4.4) gives the equality between (10.4.1) and (10.4.2). Condition (10.4.3) then follows from the Einstein field equations for the Robertson–Walker models if $\rho \sim \rho_c$.

Of course, we do not really know how different the Universe could have been, and still have evolved a partial awareness of its existence, but this discussion serves to set the cosmological coincidences on the same level as those mentioned earlier. It suggests the Universe would not have produced us but for certain accidents. In order to be self-consistent the Universe must evolve to produce us: we call this the *anthropic principle*. More picturesquely, we can say that we have climbed the mountain because we are here. This illustrates that an anthropic account does not explain the accidents of nature, but it may serve to divest them of their aura of mystery. Nor does it preclude a scientific, causal explanation.

The anthropic principle can be used to dispel our surprise at all sorts of features of the world. In fact, we expect our observation to divide into two categories: either a property of the Universe is as it is because according to our laws of physics most universes have this property and we should therefore be surprised if it were otherwise; or it is as it is because only in such circumstances could we evolve to know it. Genuine surprise should be reserved only for observations which fall into neither category.

The isotropy of the Universe can also be studied in the light of these remarks. In particular, we can enquire into the types of universe in which anisotropic perturbations remain small. One has to define what remaining small means; a suitable requirement would be that $\sigma/\theta \to 0$ at late times, so that the shear becomes dynamically unimportant, and that $\int_{t_0}^{\infty} \sigma_i \, dt$ remains finite to ensure an undistorted microwave background. It is then found that in most types of spatially homogeneous anisotropic model these conditions are not satisfied: small anisotropies tend to grow (Collins and Hawking 1973a). Consequently, if we were to live in such a universe it would be a pure accident that we now observe a small anisotropy. The remaining types, where anisotropies can die out, include those that admit Robertson–Walker models as special cases, but these types are highly exceptional. Given a random choice of anisotropic model the probability of choosing one of these special types is zero (which, for the technical reason that there are an infinity of choices and $\infty \times 0 \neq 0$, does not mean it is impossible to choose one). If these were the only universes to admit the possibility of life, an argument pursued by the cited authors, then, according to the anthropic principle, we should not be surprised to find ourselves in one of them. While one cannot claim that this satisfactorily explains the isotropy of the Universe, it might perhaps allow us to adopt a more relaxed attitude towards it.

11.   Singularities

11.1 What is a Singularity?

In the Robertson–Walker cosmologies that we have been studying, the length scale of the Universe shrinks to zero and the matter density rises to infinity at a finite time in the past. At this time physics as we know it comes to an end, and we have some sort of 'singularity'.

We can compare this situation with that of the classical point electron. As we approach the point charge the electric field rises to infinity and at the charge Maxwell's equations break down. However, we know that long before this the domain of validity of classical electrodynamics has been transcended, and we are dealing merely with the failure of an idealisation. Indeed, all we have to do is to smear the electron out a little bit, within the classical theory, and the infinite singularity disappears.

The singularities of the Robertson–Walker models are something apparently quite different. For at those points the basis of the theory we have set up is completely undermined. In particular, at a singularity there is no region, however small, in which the gravitational field can be transformed away by adopting the viewpoint of a freely falling observer. Thus the Equivalence Principle, upon which the theory is built, fails at these points. In short, singular points are not to be regarded as space–time events, and should be removed from the model. To put it more colloquially, we cannot say what was before the big bang because we cannot say what was at the big bang. The singularity denies us the resources of the language of physics, and for that reason is not an event of our physical space–time.

The existence of a singularity can be detected in terms of what happens in the non-singular part of the space–time, because trajectories disappear off of the edge of the space–time: the histories of observers following these trajectories come to an end or have a beginning for no apparent reason!

There are a few technical details that have to be added to turn this into a definition of a singularity. If we remove from a space–time a perfectly ordinary non-singular event, or set of events, thereby creating a new model for the Universe in which these events do not exist, then an observer approaching one of these events disappears off of the edge of the new model for no apparent physical reason. But this sort of singularity can be removed simply by replacing the missing events, and so does not cause us too much grief. Thus a space–time contains a genuine singularity if it cannot be extended to a non-singular space–time.

The second technicality involves the meaning of 'coming to an end', which, to avoid excessive repetition, we shall take to include 'having a beginning'. Loosely speaking, perfectly regular space–times 'come to an end' at infinity, so we consider singularities to occur only if the end of a trajectory occurs after a finite travel time (into the future or the past). If we restrict ourselves to freely falling observers then the travel time is naturally the proper time as measured by a local freely falling clock. So a minimum requirement that a space–time be free of singularities is that freely falling observers do not come to an end after a finite proper time.

In fact, examples can be given where this condition is satisfied but in which light rays come to an end after a finite 'travel time'. The travel time along a light ray cannot be taken as proper time, since this is always zero. Instead, we can employ the so-called affine parameter, λ; this is defined, up to a trivial charge of zero point and scale, by the requirement that a freely falling observer using normal coordinates, ξ^μ, finds that null geodesics satisfy the standard equation

$$\frac{d^2 \xi^\mu}{d\lambda^2} = 0$$

locally, to first order. In fact, comparison with equation (5.11.5) shows that along a time-like geodesic the proper time plays the role of an affine parameter, so this extension to light rays is a natural one. A further requirement for a non-singular space–time is therefore that light rays continue to infinite affine parameter values.

It is reasonable to argue also that a space–time should not be called singularity-free if it contains a path that can be followed by a spaceship (hence not a free fall), which comes to an end after the expenditure of a finite amount of fuel. The theory can be extended to cover this case; but the theorems which prove the existence of singularities show that the former conditions are violated anyway, so from this point of view the extension is not vital.

A space–time in which all geodesics can be continued to arbitrary affine parameter values is said to be geodesically *complete*. Our discussion can therefore be summarised by saying that a space–time will be singular if it is incomplete for time-like or null geodesics. In principle, it is relatively easy to discuss particular highly symmetric idealisations where solutions of Einstein's equations for the geometry of space–time are known explicitly. To see whether singularities occur one has to investigate whether all solutions of the geodesic equation can be continued for arbitrarily large 'travel times'.

As an example, consider the collapse of a massive spherically symmetric gas cloud under its own gravity, which might represent a star at the end of its evolution. As the cloud contracts the gravitational field near its surface grows. Therefore the acceleration of freely falling frames towards the star increases. Since these are the frames in which light is measured locally to travel with speed c in straight lines, it follows that gravity has an increasing effect on light near the surface of the cloud. As viewed by a static observer a long way from the star, the light emitted in the vicinity of the star is pulled inwards. The existence and predicted magnitude of this effect is confirmed by the observations of the bending of starlight and radio waves passing close to the Sun. If the collapse of the spherical cloud continues such that its surface area, as measured by freely falling observers, becomes less than $16\pi G^2 M^2/c^4$ the gravitational bending passes a critical value. From any point within this limiting sphere, light pulses which appear to a freely falling observer to be sent outwards are in fact bent back towards the centre. From within this sphere there is no possibility of communicating with the outside world, and so the sphere forms a horizon for observers inside it. In the space–time picture (figure 11.1) this limiting surface is called an *event horizon* (or, strictly, an *absolute event horizon*).

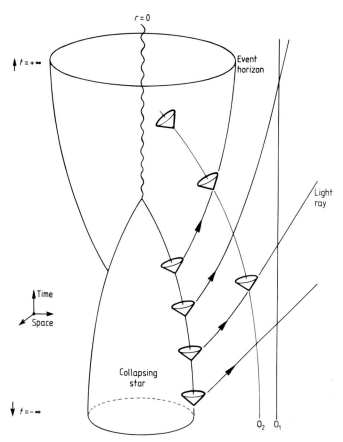

Figure 11.1. The spherically symmetric collapse of a star to a black hole. Everything inside the event horizon must fall into the singularity at $r = 0$. Outside the event horizon, an observer O_1 sees light emitted from the stellar surface increasingly redshifted since equally spaced pulses take increasing times to reach him. As the infalling observer O_2 approaches the horizon, his light cones tip over.

We *define* the 'radius', r_G, of the horizon such that its area is $4\pi r_G^2$; so

$$r_G = \frac{2GM}{c^2},$$

(11.1.1)

225

The radius r_G is called the *Schwarzschild radius* after the astronomer who first discovered this spherically symmetric solution of Einstein's equations. The object bounded by the event horizon is called a 'black hole', since it swallows everything that falls into it and lets nothing out. Note that an observer falling through the horizon notices nothing strange as he crosses it: the Equivalence Principle is valid, so local physics proceeds in accordance with special relativity.

However, the falling material of the star continues to fall towards the centre whatever might be done to try to avert this. The reason is simple: since light is being sucked inwards, any particle which does not do likewise would be travelling locally faster than light! Once inside the event horizon there is no alternative but to plunge into a singularity of infinite matter density at $r = 0$, after a finite proper time. Geodesics cannot be extended through $r = 0$. At this point, space–time comes to an end and the question of what happens 'next' is meaningless.

The 'disappearance' of matter from our Universe in this way may seem unreasonable, even though it is hidden from the sight of external observers by the event horizon. Theories that produce singularities are indeed usually considered to be unreasonable. However, at this stage it could be that the cause of the difficulty is our unreasonable assumption of exact spherical symmetry. Nothing macroscopic can be exactly spherical, and it might be conjectured that a slightly asymmetric cloud would avoid a singularity. Indeed, in Newtonian dynamics, particles in a slightly asymmetric falling cloud avoid arriving at $r = 0$ together, and can be followed through the point of maximum compression to a subsequent re-expansion. The 'singularity theorems' (Hawking and Ellis 1973) show that in general relativity this is not the case: some sort of singularity must occur even if reasonable departures from sphericity are allowed. This need not necessarily be a point of infinite density, but some geodesics must come to a premature end.

11.2. The Cosmological Singularity

The explicit solutions for dust and radiation-dominated Robertson–Walker models (Chapter 6) show these to have singularities in the past. Of course, this proves nothing about other Robertson–Walker models containing other types of matter, but in fact it is easy to show that all of these spatially homogeneous, isotropic models must be singular. For the field equation (6.6.6) is

$$\frac{\ddot{a}}{a} = -\frac{4\pi G}{3c^2}(\mu + 3p/c^2). \qquad (11.2.1)$$

If $p \geqslant 0$ and $\mu > 0$, which is fulfilled by all reasonable matter, then this shows $\ddot{a} \neq$ constant. Since $\dot{a}/a$ is observed to be positive at present, consider this case. Equation (11.2.1) then shows that

$$\ddot{a} < 0,$$

hence $a < H_0 t$, and so $a = 0$ at some time t_0 ago, where $t_0 < H_0^{-1}$.

The conservation equation (6.6.7) gives

$$\dot{\mu} = - (\mu + p/c^2)\, 3\dot{a}/a,$$

hence $\dot{\mu}/\mu \leqslant - 3\dot{a}/a$, and so $\mu \geqslant \mu_0/a^3$. Therefore $\mu \to \infty$ as $a \to 0$, and there is a real singularity a finite time in the past if $(\dot{a}/a) > 0$ at any time. If $(\dot{a}/a) < 0$ at some time there is a real singularity in the future.

It now has to be shown that the existence of singularities is not an accident of the high symmetry. To see how the argument goes we first look at the reason for the occurrence of singularities in the Robertson–Walker dust models from a different point of view. At time t_0 a sphere of radius $a_0 R$ will contain a mass $(4\pi/3)\rho_0 a_0^3 R^3$. This sphere will be smaller than its Schwarzschild radius if $a_0 R$ is less than $(8\pi G \rho_0 a_0^3 R^3)/3c^2$ (equation 11.1.1); hence, if

$$a_0 R > \left(\frac{3c^2}{8\pi \rho_0 G}\right)^{1/2} = \left(\frac{q_0}{2}\right)^{1/2} \frac{c}{q_0 H_0}; \qquad (11.2.2)$$

i.e. if $R > (q_0/2)^{1/2} R_\infty$ (equation 8.9.1). In the time-reversed contracting model, the condition (11.2.2) ensures that light rays reconverge, as in the black-hole case, and this heralds a singularity. If the real Universe is not too different from a Robertson–Walker model, the mass inside a large sphere will not differ too much from the estimate leading to equation (11.2.2), so the reconvergence of light rays should not be too much affected. Clearly we expect that singularity theorems that show the appearance of singularities as a general feature of collapsing clouds should imply the existence of a singularity in the Universe.

There are two different arguments that achieve this end, starting from slightly different assumptions and yielding slightly different conclusions. In the first version it is assumed that the microwave background has been at least partially thermalised by repeated scattering. This must be the case unless it is produced by a large number of discrete sources at large redshifts. At an optical depth of about 0.2 there is sufficient matter to ensure that null rays traced back in time will start to reconverge. This leads to the conclusion that there is a singularity somewhere in the Universe. Note that the argument depends on the thermal spectrum of the background radiation, not on the assumption that the Universe is closely represented by a Robertson–Walker model. Although we expect the singularity to have occurred in our past, rigorously speaking, this does not follow from the theorem.

Alternatively, from the assumption that the Universe is not too different from a Robertson–Walker model at recent times, it can be shown that a singularity must have occurred in the past. The assumption is justified since the microwave isotropy shows that the Universe is isotropic to a high precision back to the last-scattering surface. The argument then depends on showing that time-like geodesics traced backwards into the past start to reconverge before that. If the Universe has a high matter density $(\gtrsim \rho_c)$ with an ionised intergalactic medium, it turns out that there is enough matter to produce the required reconvergence of time-like geodesics by an optical depth of 0.5. In the opposite extreme, in a low-density universe, $q_0 \sim 0.01$, the last-scattering surface is at $z \sim 10^3$, and there is enough energy in the background radiation to produce reconvergence by about $z \sim 300$. In all cases the singularity theorems guarantee the existence of a singularity in the past.

We expect that as long as the Universe is homogeneous on a scale equal to the Schwarzschild radius given by (11.2.2), there will exist a singularity to the past of all events. Therefore, if we believe that the Universe must be isotropic back to neutrino decoupling at $t \sim 1\,\mathrm{s}$, then all events later than one second have a singularity in their past. If the Universe were highly anisotropic before this time then, so far as has been shown, there could exist regions with non-singular pasts.

11.3. Bangs and Whimpers

These arguments prove only that freely falling 'observers' have an untimely beginning in spontaneous generation. They do not prove that the Universe must have originated from a state of infinite density. In the case of spatially homogeneous models it is possible to discuss in some detail the types of singularity that can occur.

To begin with we look at the behaviour of the trajectories of the observers in these models who 'see' their universes as spatially homogeneous at any time. In the isotropic models these fundamental observers are attached to freely falling clusters of galaxies. In the anisotropic models this is not necessarily the case. They need not follow the paths of any of the matter in the models and they need not be in free fall. Nevertheless we can consider how their trajectories diverge and how this divergence must be controlled by the gravitational effects of matter. Let $l(t)$ be the scale length defined by these observers, so $\frac{4}{3}\pi l^3$ is the spatial volume occupied by a given set of fundamental observers at cosmic time t. The Einstein field equations yield an expression for the rate of expansion of the volume of this comoving region of the form

$$3\ddot{l}/l = -4\pi G/c^2(\bar{\mu} + 3\bar{p}) - 2\bar{\sigma}^2, \tag{11.3.1}$$

where $\bar{\mu}$ and $\bar{p}$ are the mass–energy density and pressure measured locally by the fundamental observers, and $\bar{\sigma}$ is the rate of shear of their trajectories (Chapter 9).

Equation (11.3.1) bears a strong resemblance to equation (11.2.1). Indeed, we can take over the argument used in §11.2 to prove that $l(t) \to 0$ as t tends to some finite time t_0 in the past. For $\bar{\sigma}^2$ merely acts as a further positive contribution to $\bar{\mu} + 3\bar{p} > 0$, tending to reconverge the trajectories in the past. This does not quite prove that a big-bang singularity must occur, for it may not be possible to take the solution back this far. One could conceivably have $\bar{\sigma}^2 \to \infty$, $\ddot{l} \to -\infty$ as $l \to l_0 > 0$; in such a case we should have a singularity (in the sense of incomplete geodesics), although the matter density and pressure were everywhere finite. In general, all we can conclude is that the model must cease to be spatially homogeneous before we get back to t_0, if it is not to develop a singularity at t_0, where the trajectories of fundamental observers intersect (figure 11.2).

228

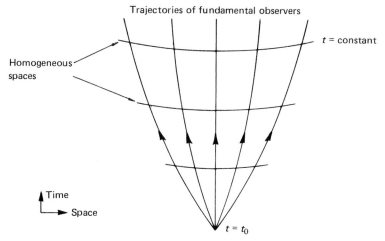

Figure 11.2. Space–time diagram of paths of fundamental observers at rest relative to the surfaces of homogeneity. The paths intersect at $t = t_0$.

There are two cases to consider. The simplest is the case in which the fundamental observers do in fact follow the trajectories of freely falling material particles, despite the presence of anisotropy. Then $\bar{\mu} = \mu$, $\bar{p} = p$, and $l = a$. In this case it can be shown that the space–time *can* be continued back to t_0, where $a \to 0$, so just as before, we must have $\mu \to \infty$. Thus there is an infinite-density singularity (a big bang), although, as discussed in §9.2, of possibly a different type from that in the Robertson–Walker models.

In the alternative case the matter does not flow along with the fundamental observers. This may be due to rotation of the matter, or to a 'tilt' of the homogeneous surfaces with respect to the matter (figure 11.3). So-called 'tilted universes'

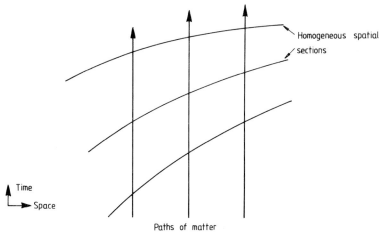

Figure 11.3. A *tilted* universe. An observer flowing with the matter has a net velocity relative to observers for whom space is homogeneous.

229

are particularly interesting because some of them are amenable to calculation (Collins and Ellis 1979). It is known that a new type of singularity may occur in tilted universes, and examples are known explicitly in which it does. Here $\mu \neq \bar{\mu}$ because the two quantities are measured by different observers. At the singularity μ remains finite; but an observer freely falling into (or emerging out of) the singularity measures $\bar{\mu} \to \infty$, essentially because the relative velocity of the two classes of observers approaches the speed of light there. The trajectories of these freely falling observers cannot be extended, so these models have genuine singularities, but of a milder type than a big bang; they have been called 'whimpers'. As a corollary to the preceding discussion, a breakdown of spatial homogeneity must occur in these models.

The distinction between whimpers and bangs can be illustrated in a special relativistic context (Ellis 1975). We relinquish the connection between the matter flowing in the space–time and the space–time geometry that is provided by the Einstein equations, and imagine that we can specify each independently. This allows us the simplicity of specifying the geometry to be that of Minkowski space–time. For the purposes of illustration we also drop the requirement that the matter flow should be induced by a realistic pressure distribution and allow the world-lines of matter to be given arbitrarily. Furthermore, we restrict ourselves to two dimensions.

In Minkowski space–time the surfaces (strictly, lines in two dimensions) $x^2 - t^2 = $ constant are homogeneous in the sense that by Lorentz transformation we can move any point to any other point in the same surface. Furthermore, if the constant is positive the surfaces are space-like, and hence spatially homogeneous (figure 11.4). In region II of Minkowski space–time, the appropriate fundamental observers follow straight lines through the origin. In this region we may add matter flowing with sufficient symmetry that the matter density is constant on the homogeneous surfaces, hence that the universe looks homogeneous to the fundamental observers. It is obvious that a big-bang singularity occurs at O.

By contrast, consider regions I and II of Minkowski space–time with the trajectories of particles added in a rather different way (figure 11.5). Again, this can be made into a spatially homogeneous model in region II, even though it may not look so, essentially because of the way the Lorentz transformation works. Region I, however, is not spatially homogeneous. The matter world-lines 'pile up' as one approaches surface S, which is a whimper singularity. To see this, note that an imaginary observer freely falling (back in time) into the singularity at O sees an infinite amount of matter in a finite time, so $\bar{\mu} \to \infty$; while μ has a finite value, equal to its constant value on H, which is a homogeneous (null) surface.

A further example of a whimper singularity is provided by the steady state theory. Here, instead of adding world-lines of matter to Minkowski space–time, one adds them to the de Sitter model (§6.1). A two-dimensional picture of de Sitter space–time is provided by the surface of a hyperboloid (figure 11.6). The steady state model is one-half of this hyperboloid obtained by slicing symmetrically with

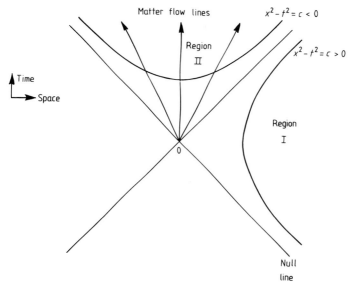

Figure 11.4. Space-like (in region II) and time-like (in region I) surfaces (lines) of homogeneity in two-dimensional Minkowski space–time. Regions I and II are bounded by light rays through P. By the addition of matter, region II becomes a spatially homogeneous universe with a big bang at O.

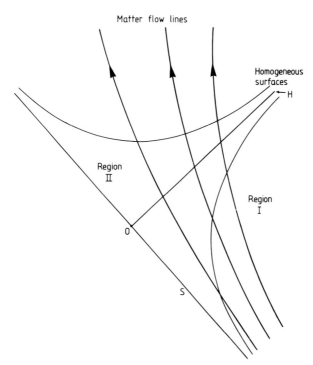

Figure 11.5. A (two-dimensional) 'whimper' model.

231

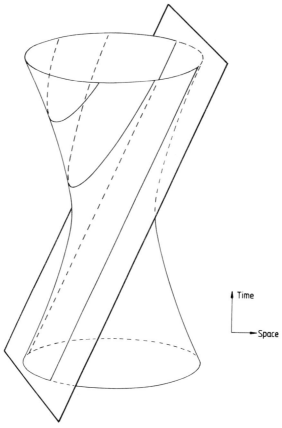

Time

Space

Figure 11.6. The surface of the hyperboloid prepresents the geometry of de Sitter space–time (with two dimensions suppressed). The de Sitter model contains no matter; the curvature arises solely from the effect of a non-zero cosmological constant. The upper half of the surface obtained by slicing the hyperboloid with a plane through the centre is the space–time of the steady state theory.

a plane through the origin. This is opened out in figure 11.7; the bounding surface S is a whimper singularity. To see this note that the matter density μ is constant everywhere, yet a freely falling observer travels back to the boundary in a time which is finite. This follows because his geodesic is extendable in the full de Sitter space–time, even though it is not extendable in the steady state region. Since this observer passes an infinite amount of matter on the way, he must find $\bar{\mu} \to \infty$.

Whimpers can be shown to occur also in the full theory of general relativity, when the connection between the matter flow and the geometry is restored, although this may depend on the assumption of exact spatial homogeneity. However, a slight departure from homogeneity need not necessarily produce a big bang; we could have 'little bangs' in which only *some* of the matter content of the model universe originates in an infinite-density singularity. Nevertheless, whimper singularities have been shown to be very special cases even within the restricted class of

232

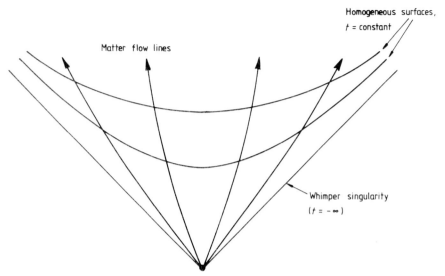

Figure 11.7. The steady state space–time. The top half of the hyperboloid of figure 11.6 has been opened out and matter flow lines added. The homogeneous surfaces correspond to the curves on the hyperboloid of figure 11.6. The boundary (at $t = -\infty$) is a whimper singularity.

exactly spatially homogeneous models. *A priori* this means they are extremely unlikely to occur (they have zero probability in the sense noted in §10.4). One should, however, resist the temptation to draw too strong a conclusion from this. For example, black holes are extremely special configurations which have zero probability of occurring in this technical sense, yet, according to the theory, they are the inevitable end-point of the evolution of massive stars!

Whatever the eventual outcome of these considerations, one might expect it to be of little consequence for that part of the evolution of the Universe that we think we do understand. However, there is one important point to be made. Consider again the discussion of the singular point electron with which we began. The problem is not really resolved by smearing the charge out by *fiat*; this would actually be inconsistent with the classical theory, according to which a smeared out electron would fly apart. What really happens is that quantum theory comes to the rescue, providing us with a way of formulating physical laws on scales at which the classical theory is demonstrably inapplicable. It is tempting to argue that the break-down of physical laws at a space–time singularity is not serious because before the singularity can be reached the density of matter will have risen to the point where a classical theory of gravity is clearly inadequate anyway. Quantum theory should be introduced at this stage with as yet unknown consequences. The existence of whimper singularities, at which the matter density is finite, shows that, on the contrary, one cannot simply invoke quantum mechanics to save general relativity from consuming its own premises.

233

12.   The Evolution of Structure

12.1.   The Problem of Structure

As a consequence of the existence of small-scale inhomogeneity the Universe is now in a state of gross disequilibrium. Yet it appears to have begun in equilibrium and will end in equilibrium. One cannot help but feel that it might have chosen an easier, if less interesting journey between its beginning and its end. The problem of structure presents us with the task of accounting for the development of this inhomogeneity.

The obvious approach to the problem is to seek to show that the evolution is from an *unstable* equilibrium to a stable one. Small initial fluctuations, which are natural in a sense to be specified by the theory, might be amplified to produce the observed structure. Or one can imagine the appearance of order through a process akin to the spontaneous breaking of symmetry that leads, for example, to the domain structure of ferromagnetism, or by a process analogous to a phase transition in which finite fluctuations condense out of a uniform medium. In the progression from a smooth, small-scale universe to ordered macroscopic structures, all of these mechanisms are to be seen at work. Therefore it would not be surprising to see any or all of them operating in the development of substructure from a universe uniform on the largest scale. However, from the point of view of theoretical tractability, it is easier if one remains close to a homogeneous universe. This is possible, even for finite perturbations, as long as condition (9.1.1) is satisfied. It is therefore for this case that most of the calculations have been developed, and therefore with the fate of initial fluctuations that most theories deal.

In §12.2 we start with the problem of matter–antimatter asymmetry. It would clearly be elegant if the Universe were to be initially matter–antimatter symmetric, and for the separation to occur as part of the development of structure, but it appears that this cannot be so. Subsequent sections are devoted to a brief review of the problems and theories of galaxy formation, and in the final section we consider the fate of the Universe and the arrow of time.

12.2.   Matter–Antimatter Symmetric Universes

The rate of annihilation of nucleon–antinucleon pairs in the early Universe is $10^{-11} n_N$ s^{-1} for a nucleon number density n_N m^{-3}. This leads to annihilation times

234

much less than the expansion timescale below about 10^{12} K, when the nucleon pairs cease to be in equilibrium. The annihilation rate depends on a cross section that can be measured in the laboratory at the relevant energies, so cannot be substantially wrong. It follows that most of the nucleon–antinucleon pairs annihilate in the early stages. One might argue that in fact we need only a small number left over (about one in 10^8-10^9) to make the material universe, so that these might be the remnants from a symmetric big bang, containing equal numbers of particles and antiparticles. However, more detailed calculations of the equilibrium between nucleon pairs and photons in the early Universe show that a particle–antiparticle symmetric and spatially homogeneous big bang could have no more than 10^{-18} nucleons for every photon in the present-day Universe.

There are two ways one might try to get round this. First, the Universe cannot have been exactly homogeneous. There must have been statistical fluctuations, ΔN, in the difference $N_n - N_{\bar{n}}$ between the number of nucleons and antinucleons in a given region, even if this difference is zero on average. For small statistical fluctuations, if $\Delta N \ll N_n$, we have

$$\frac{\Delta N}{N_n} \sim N_n^{-1/2}.$$

as in §3.3. But then ΔN nucleons survive in this region while $N_{\bar{n}} \sim N_n$ annihilate to give the microwave background. Thus $N_n \approx N_\gamma$ and $\Delta N \sim N_n^{1/2} \sim N_\gamma/\Delta N \sim 10^8$-$10^9$. Therefore statistical fluctuations leave regions of separated matter and antimatter of only 10^8 particles, or a mere 10^{-19} kg!

The second way one might seek to maintain a symmetric universe is to alter the physics in the early stages in such a way that particles and antiparticles do not get together to annihilate. Essentially one is asking here for large non-statistical fluctuations – this is not an impossible request. In particular, Omnès (1969) has claimed that in the Hagedorn theory (§6.9) of strong interactions there is a phase transition at a temperature of about 10^{12} K. Above this, regions of matter and antimatter tend to be separated, mixing only as a result of diffusion. At lower temperatures, nucleons and antinucleons then mix and annihilate as normal. Unfortunately, a typical region of matter contains only $N_n \sim 10^{27}$ nucleons or 1 kg at the critical temperature. It is possible that this is exactly right to allow the build-up of large separated regions of matter and antimatter by subsequent coalescence without too much annihilation. But clearly the scheme loses some of its initial attractiveness at this point.

Recent advances in elementary particle physics have, however, set the problem in a new light. In particular, there now exist viable theories, in the sense that they are not contradicted by experimental evidence, that unite the strong, weak and electromagnetic interactions. These *grand unified theories* (GUTS) have in common the feature that baryons are not conserved. Thus, for example, the proton is predicted to be unstable (although with a very long half-life). Of more relevance to cosmology is the possibility that from a non-equilibrium particle–antiparticle symmetric initial state at 10^{-42} s, an equilibrium state with an excess number of particles may emerge at 10^{-5} s. The result depends on the prediction of the theory that whereas under normal conditions the baryon non-conserving interactions are small, at temperatures in excess of 10^{28} K, which occur before 10^{-36} s, they have the same strength as the other forces. Suppose then that a certain particle in the theory, and its antiparticle, decay and annihilate too slowly to remain in thermal equilibrium at a time when baryon non-conserving interactions are still strong. If the particle and antiparticle can decay subsequently into states with different numbers of baryons and antibaryons, and if they undergo their respective decays at different rates, an excess of baryons can be built up. There is a bonus as well — along with the particles, these theories generate a lot of entropy. For a suitable choice of parameters they can produce the required 10^9 photons per baryon! (see Turner and Schramm 1979).

The weight of theoretical and observational evidence (§4.15) therefore points to a tentative conclusion that after 10^{-5} s the material Universe is not matter–anti-matter symmetric. But the grand unified theories suggest that high-energy physics, operating at the only time when the highest energies were available, may be able to provide an explanation.

12.3. The Growth of Inhomogeneity

Suppose a small irregularity occurs in a uniform static self-gravitating gas. If pressure forces are negligible an overdense region will tend to increase its density further as a result of its increased gravitational attraction, and will continue to grow until pressure forces do intervene. The gravitational field at each point, ϕ, is given by Poisson's equation

$$\nabla^2\phi = 4\pi G\rho,$$

while the velocity of matter, $\mathbf{u}$, is controlled by the Euler equation of motion,

$$\frac{\partial \mathbf{u}}{\partial t} + \mathbf{u}\cdot\nabla\mathbf{u} = -\nabla\phi. \tag{12.3.1}$$

and the conservation equation for the density, ρ,

$$\frac{\partial \rho}{\partial t} + \nabla\cdot\rho\mathbf{u} = 0. \tag{12.3.2}$$

To investigate the development of a perturbation in the density, $\rho' = \rho - \rho_0$, and velocity $\mathbf{u}'$, from a uniform static background, density ρ_0, we linearise these equations in the perturbations ρ' and $\mathbf{u}'$. Eliminating $\mathbf{u}'$, we obtain

$$\ddot{\rho}' = 4\pi G\rho_0\rho'. \tag{12.3.3}$$

From equation (12.3.3) we deduce that the density perturbation grows exponentially on a timescale $(4\pi G\rho_0)^{-1/2}$. The system is unstable, and any naturally occurring fluctuations would grow to a finite size in a reasonable time.

When a corresponding analysis is performed for the fate of density fluctuations in a uniform expanding universe, the result is significantly different.

To see this we take the simplified Swiss cheese model of §9.5 with zero pressure. Inside a given spherical region we imagine the mass to be uniform, but concentrated slightly, so that it behaves as a universe with a higher mean density (figure 12.1). Labelling the background universe with subscript zero, and the

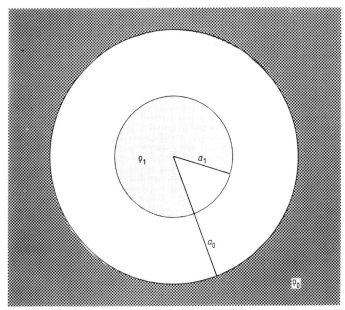

Figure 12.1. A spherical perturbation (density ρ_1, radius a_1) in a hole (radius a_0) in a Robertson–Walker universe of density ρ_0.

perturbed universe with subscript 1, we have

$$\rho_0 a_0^3 = \rho_1 a_1^3 = 3M/4\pi = \text{constant}$$

in time. This follows from the conservation of mass in the perturbed region and the definition of the Swiss cheese model. The energy equations (6.6.4) are

$$\tfrac{1}{2}\dot{a}_i^2 - \frac{GM}{a_i} = E_i = \text{constant}, \qquad (12.3.4)$$

with $i = 0, 1$. If we set

$$\delta = a_1/a_0 - 1,$$

and assume $\delta \ll 1$, then to first order in δ, the equations of motion obtained by

237

differentiation of (12.3.4) give

$$\ddot{\delta} + 2\frac{\dot{a}_0}{a_0}\dot{\delta} = 3\delta\frac{GM}{a_0^3}.$$

As usual, to get a quantitative estimate, take the Einstein–de Sitter model with $a_0 = (9GM/2c^2)^{1/3}t^{2/3}$ (equation 6.7.7). Then

$$\delta = At^{+2/3} + Bt^{-1}.$$

The relatively decreasing mode Bt^{-1} is unimportant here. The growth of the increasing mode is not exponential as before, but a power law. The form of this result is valid for other models and arbitrary perturbations. Originally the suppression of exponential growth was interpreted to mean that finite fluctuations could not form out of infinitesimal ones in a finite time, and hence that the Universe was insufficiently unstable to grow galactic structure in a natural way. In fact, of course, one can draw the alternative conclusion that provided one starts early enough any desired amount of inhomogeneity can be obtained. For example, statistical $N^{1/2}$ fluctuations at 10^{-34} seconds will amplify to galaxy-sized density contrasts, $\delta\rho/\rho \sim 1$, at $z \sim 10$.

The real problem raised by this discussion is how one can use it to explain anything, for suitable initial irregularities will yield desired final states, and apparently 'reasonable' initial states (such as statistical fluctuations at the Planck time or Compton time) produce totally unreasonable outcomes. Clearly what is needed is some natural way of explaining why irregularities grow to give just the features of galaxies. This could involve providing an initial spectrum of fluctuations of the right sort at the start of the era of classical amplification; or it could involve selecting preferred structures by physical processes at a later stage. In either case we have to produce galaxies with a characteristic mass, M_*, associated, presumably, with their characteristic luminosity L_* (Chapter 1), and we have to account for their characteristic size or density. The theory should also account for their spin and shape and clustering properties. The key point appears to be that if all of this is to be obtained without simply including it in the initial conditions, then some force in addition to gravity must be operating, because gravity provides no characteristic scale.

To illustrate the point, consider the addition of pressure forces to the above discussion in the non-expanding case. Then equation (12.3.1) acquires an additional term $-1/\rho\nabla p$ on the right-hand side. In the adiabatic case, $p \propto \rho^\gamma$, equation (12.3.3) becomes

$$\ddot{\rho}' = 4\pi G\rho_0\rho' + c_s^2\nabla^2\rho'$$

to first order in ρ', where $c_s = (\gamma\rho_0/\rho_0)^{1/2}$ is the adiabatic sound speed. If we look for wave-like solutions of this equation, of the form $\rho' \propto \exp i(\omega t + \mathbf{k}\cdot\mathbf{x})$, we obtain the dispersion relation

$$\omega^2 = -4\pi G\rho_0 + c_s^2|\mathbf{k}|^2.$$

For $c_s^2|\mathbf{k}|^2 < 4\pi G\rho_0$, we have $\omega^2 < 0$, hence exponential growth and no propagation. Pressure therefore acts as a stabilising force on scales such that $c_s^2|\mathbf{k}|^2 > 4\pi G\rho_0$, or

$$l = \frac{2\pi}{|\mathbf{k}|} < \left(\frac{\pi}{G\rho_0}\right)^{1/2} c_s.$$

On larger scales, gravity dominates and the perturbations are unstable. The mass in a region of the critical scale is called the Jeans mass, M_J, and is

$$M_J = \tfrac{4}{3}\pi^{5/2} \frac{c_s^3}{(G\rho)^{3/2}}.$$

The way in which galaxy generation might now be expected to operate is illustrated in Hoyle's original theory (1953). Consider a large diffuse cloud of galactic mass destined to become a galaxy. As it collapses under its own gravity it heats up, stabilising at around 10^4 K because at this point cooling is particularly efficient. Thereafter, c_s is constant as ρ increases. Thus M_J decreases, and so the scales of fluctuations which are unstable decrease. The cloud fragments, the pieces collapse and fragment, and this continues until optically thick clouds are formed. These will become stars.

Of course, the origin of the pregalactic cloud is not accounted for in the theory. One might hope that a similar type of process operating on a larger scale could be responsible. However, while at any stage the growth of large-scale perturbations is favoured $(M \geqslant M_J)$, as the universe evolves this encompasses all scales, and there is no obvious way of generating a characteristic galactic mass, except by imposing this on the initial spectrum of fluctuations.

The existence of the microwave background alters this picture in an important way. While the radiation is strongly coupled to matter, the above discussion is not valid. The radiation produces a viscous drag on the electrons, thereby preventing all perturbations from growing, and actually dissipating some. One can use this either as a way of generating the desired characteristic scales at the recombination epoch from a reasonably smooth input at $t \sim 10^{-5}$ s, or as a way of generating a particular spectrum of fluctuations upon which later physical processes will impose the required characteristic pattern.

Instead of making all structure by breaking things up one can imagine it appearing also as a result of the clustering and possible coalescence of objects under their mutual gravitational attraction. The various theories differ in the importance attached to this, and in the scale on which it is supposed to operate. The absence of characteristic length scales over a large range of clustering scales is suggestive of the action of gravity, but by no means conclusive.

To summarise, the broad picture is this. An initial type of fluctuation is postulated, perhaps as a result of physical considerations, perhaps on grounds of simplicity. The fate of these fluctuations is followed through the radiation-dominated era and thence up to recombination. This leaves inhomogeneities which will either fragment or cluster to give the pattern we now have. If dissipation processes are invoked to speed things up the final formation of galaxies can occur later than $z \sim 10$, and may even still be occurring now, without contradicting (the lack of) observational evidence. Otherwise, galaxy formation must be complete at an earlier epoch.

Even within this broad outline many variants are conceivable. For example, proto-globular clusters could undergo collapse and fragmentation into stars, while at the same time clustering into galaxies. However, not all conceivable options are physically permitted. In particular, the choice between the fragmentation picture and the clustering picture depends on the characteristic mass of the perturbations at the time of recombination, which is itself governed by the type of initial fluctuation (§12.4). There are also theories which lie outside this broad scheme; for example, that matter agglomerates around lagging remnants of an inhomogeneous big bang.

12.4. The Input of Structure

There are two key features of the initial fluctuations which determine their fate. These are the distribution of matter on various scales and the relation between the distribution of matter and of radiation.

The distribution of matter is measured by the way in which the excess density of a region scales with the total mass of the region. It is usual to consider power law distributions of the form

$$\delta\rho/\rho \propto M^{-\alpha}. \tag{12.4.1}$$

For example, $\alpha = \frac{1}{2}$ corresponds to random statistical fluctuations, $\delta N/N \propto N^{-1/2}$. A measure of the associated gravitational perturbation is $\delta\phi/\phi$ for the visible universe. Thus, for a perturbation of length scale l,

$$\frac{\delta\phi}{\phi} \sim \frac{G\delta M}{l} \bigg/ \frac{GM}{(c/H)} \propto \frac{\delta\rho}{\rho} l^2.$$

It follows that $\alpha = \frac{2}{3}$ gives a particularly simple fluctuation spectrum. For then $\delta\phi/\phi = $ constant in time as each scale fills the observable universe, since $M \propto l^3$. Consequently, if $\delta\phi/\phi$ is small at one time, we can treat the universe as a perturbed homogeneous universe at all times. For $\alpha > \frac{2}{3}$ the universe is initially *more* inhomogeneous than it is now, and *ipso facto* a chaotic initial state is being envisaged. If $\alpha < \frac{2}{3}$ the universe will ultimately depart from approximate homogeneity, since it will look increasingly irregular as we see larger scales. Of course, in either case there could be some cut-off which invalidates these conclusions. Nevertheless, the $\alpha = \frac{2}{3}$ case is clearly the simplest postulate in a state of complete ignorance.

With regard to the role of the radiation, two main types of density perturbation are possible: the matter can be perturbed without the radiation; or both can be perturbed together. In the former case we speak of isothermal perturbations, and in the latter of adiabatic perturbations. The latter term arises because the entropy

240

per baryon is unaltered by the perturbation in this case. In principle, the radiation can be perturbed and not the matter, but this possibility appears to be of no importance. Even if there is no reason to expect the perturbation to be of exactly one type, one might hope that the actual behaviour would be more like one than the other. Massive neutrinos, if they exist, could also be an important source of density perturbations because they decouple from the radiation at an early stage.

An alternative proposal envisages that the matter had an initial vorticity, so that the universe started from a turbulent state of which the present rotation of galaxies is the memory. Although at first sight this appears to be a rather complicated hypothesis, putting somewhat more of what is to be explained into the initial conditions, it was originally claimed to be a simpler assumption. For the turbulent velocities remain constant in time in the radiation era. Therefore only their amplitude and not an initial time need be postulated. In contrast, a density perturbation must be given as an amplitude at some initial time (but see §12.5).

Since we are setting initial conditions, there is no upper limit to the scale of the fluctuations. However, one can distinguish between fluctuations which fit into the visible universe at the appropriate time and those which extend beyond the horizon. The former are called spontaneous and the latter primordial. Spontaneous fluctuations could have arisen from the operation of physical processes over causally connected regions. Primordial fluctuations must exist *ab initio*. All theories require primordial fluctuations which are then amplified by the operation of physical processes (see, for example, Jones 1976).

The spectrum and type of primordial fluctuations are presumably determined by physical processes operating at an earlier stage. Of particular interest in this respect are recent grand unified theories of the strong, weak and electromagnetic interactions (§12.2). Admittedly tentative calculations suggest that only adiabatic perturbations can survive when the Universe has cooled to a temperature of 10^{31} K, because at higher temperatures baryon non-conserving reactions operate to eliminate regions of below average entropy per baryon. This would lend support to the fragmentation picture of galaxy formation (§12.5).

Current theories of the amplification of perturbations to the present density contrast require $\delta\rho/\rho \sim 10^{-4}$-$10^{-3}$ at the time when the length scale under consideration first enters the horizon of the visible universe. This is not only quite large; it also suggests a relation between scale and amplitude. There are two ways one might conceive of to achieve this: either through nonlinear classical processes or by quantum mechanics. At the classical level, one can envisage a phase transition such that the equilibrium state at earliest times would not be homogeneous, but would contain some definite spectrum of inhomogeneity. In quantum mechanics the Uncertainty Principle links energy with frequency (or wavelength), and the Planck mass provides a natural amplitude $[(hc/G)^{1/2} \sim 10^{-8} \text{kg}]$ in the quantum gravity era at the Planck time $[(Gh/c^5)^{1/2} \sim 10^{-43} \text{s}]$. Indeed, with power law amplification an $\alpha = \frac{2}{3}$ fluctuation spectrum at this time produces $\delta\rho/\rho \sim 1$ on galactic scales by $z \sim 10$.

12.5. The Emergence of Structure

Suppose we start from a homogeneous universe modified by small isothermal perturbations. While the radiation and matter are coupled together, radiative viscosity prevents both the growth and dissipation of these fluctuations. The characteristic mass of perturbations that can evolve to give the observed inhomogeneity is therefore the smallest mass that becomes unstable after recombination. This is the Jeans mass for matter, $M_J \sim 10^5 q_0^{-1/2} M_\odot$. Since this is of the order of the mass of a globular cluster, isothermal fluctuations are associated with the *gravitational instability* picture of galaxy formation, in which larger-scale structure is built up by clustering and aggregation of these subunits (Fall 1979). The production of globular clusters in this way would explain their constancy of character. The main support for the scheme comes from the correlation analysis of galaxy clustering, in particular the absence of a preferred length scale. Fluctuations with $\alpha = \frac{2}{3}$ (equation 12.4.1) yield a spectrum at recombination for the *matter* distribution which is $(\delta\rho/\rho) \propto M^{-1/3}$, in agreement with the result $(\delta\rho/\rho) \sim (M/10^5 M_\odot)^{-1/3}$ deduced from a correlation function analysis. On the other hand, there is some dispute as to whether computer simulations of the clustering of point masses under their mutual gravitational attraction in an expanding universe adequately reproduce the observed structure. In particular, objectors point to evidence of a cellular pattern in the galaxy distribution on supercluster scales (see Longair and Einasto 1978).

In an alternative version of this scheme the amplitude of the fluctuations is envisaged to grow to the extent that they cannot be regarded as small perturbations, and nonlinear effects become important. Since the dynamics is controlled by the density of radiation, large *matter* perturbations $(\delta\rho/\rho \gg 1)$ can exist without affecting the homogeneity of the model. These can develop to produce a pregalactic epoch of star formation. Supermassive stars would evolve rapidly to supernovae, thereby synthesising heavy elements and returning the metal enriched material to the pregalactic medium.

In what is currently the main rival route to galactic structure, the initial input is taken to be adiabatic fluctuations. In this case radiative viscosity prevents the growth of inhomogeneity prior to decoupling, while small-amplitude perturbations can be damped out by the diffusion of photons out of regions of higher density. The characteristic mass available to produce galaxies is that corresponding to the distance through which photons can diffuse prior to recombination. This is $M_D \sim 10^{12} q_0^{-5/4} M_\odot$. At recombination, smaller-scale perturbations of lower mass have been damped out. The mass M_D is of the order of that of clusters of galaxies, so adiabatic fluctuations are associated with the *fragmentation* picture of galaxy formation. Support for this theory is based on claims that it can produce galaxies with the correct general features (e.g. Binney 1977, Silk 1977).

If the perturbations become large enough to introduce nonlinear effects, shock waves may form and lead to further damping on small scales. The characteristic mass of

242

a surviving perturbation is the Jeans mass for matter and radiation at recombination, $M_J \sim 10^{15} q_0^{-2} M_\odot$. However, the spectrum turns out to be $(\delta\rho/\rho) \propto (M/M_J)^{1/3}$ in contradiction with observations. Another ingenious variant envisages the nonlinear interaction of adiabatic waves leading to isothermal perturbations!

Alternatively, relatively large adiabatic fluctuations, of the order of a few per cent when they enter the horizon, can collapse to form black holes in the early universe. In a hot big bang, under special conditions, these can be produced in the right numbers and masses to form the 'seeds' for galaxy formation (e.g. Barrow and Carr 1978).

An attractive feature of this idea that the most reasonable way of powering active galaxies and quasars is by means of central black holes having masses of order $10^8 M_\odot$. Gas falling into a hole can convert its gravitational potential energy into radiant energy in one of a number of possible ways, and only a reasonable amount of infall is needed to power a quasar if the black hole has a mass this high. In addition, although this is much more speculative, the point source of radio emission in the centre of our own Galaxy could be a black hole which is no longer being fed with much gas. In that case one would expect all galaxies to possess central black holes, and perhaps to have spent part of their lives in the quasar phase. Certainly the proportion of active galaxies (about 1%) and their lifetimes, as estimated from the separation of the radio lobes of giant double radio galaxies, are consistent with this hypothesis. A simple picture would then be that galaxies form from the material accreting round super-massive black holes by fragmentation of the inflowing gas.

Unfortunately it appears that very special conditions must be postulated in the initial phases if just the right number of black holes are to be formed, which means that the scheme is rather improbable. An alternative possibility is a cold initial big bang, in which the microwave background itself would be produced from accretion energy in the early stages. In this picture the holes would not be so massive ($\sim 10^6 M_\odot$), but could form over a range of physical conditions. Some clustering of the holes initially would be expected, so galaxies would form around seeds of several black holes which would now form halos of dark matter and could provide the missing mass in rich clusters!

In the case of both adiabatic and isothermal fluctuations velocities are easily damped, so galactic rotation must be produced by some form of tidal interaction. The problem here appears to be the difficulty of producing sufficient rotation. In this respect it is pertinent to note that there is evidence that the rotation rate of elliptical galaxies is much less than that suggested by their ellipticity.

The idea that the rotation of galaxies might be a relic of a turbulent origin to the Universe is an appealing one. In a stationary turbulent fluid, nonlinear interactions lead to a unique distribution of velocity amongst an intermediate range of eddy scales. This is the Kolmogorov spectrum

$$v(l) \propto l^{1/3}$$

for the velocity associated with an eddy scale l. There is also a scale, $l_c = vt_{exp}$ at a cosmic time $t \sim t_{exp}$, below which the turbulence will be damped by radiative viscosity. The maximum scale, obtained at t_{eq} (§6.8), is $v_{eq} t_{eq}$. For a reasonable choice of the single free parameter v_{eq}, the characteristic mass associated with this scale is about that of a galaxy. Furthermore, one obtains $\delta\rho/\rho \propto M^{-2/3}$ for a range of larger masses.

The theory has been criticised, however, on various grounds. In this scheme the universe cannot be nearly homogeneous at early times. While this is not in itself a condemnation, it does appear to destroy some of the simplicity, since it seems that more than one parameter must in fact be specified. It may lead also to problems with light element abundances (compare §9.3). In addition, there are energy problems, since the turbulence must be maintained against dissipation on small scales by energy input on large scales. It has been claimed that this destroys the agreement of the characteristic mass in the theory with the characteristic galaxy mass for any choice of input parameters.

Finally, recall that the fluctuations at recombination may not indicate the final characteristic mass scales. They could provide only the input on which later physical processes might work. In particular, galaxies could be formed at comparatively recent epochs ($z < 10$) with the characteristic scale determined not by what is there to collapse and fragment, but what can actually do so. The point here is that in order to collapse, a cloud must be able to cool in a timescale less than the expansion time. This appears to yield a characteristic galactic mass in clouds of supercluster scale (e.g. Rees and Ostriker 1977).

12.6. The Fate of the Universe and the Arrow of Time

The Universe appears to evolve from a past to a future. The formation of structure involves an increase in entropy, and its dissolution is an aspect of the evolution of physical systems towards equilibrium. We observe this *arrow of time* despite the fact that all of our physical laws are reversible in time. Thus, for every system that evolves towards equilibrium we can consider another system in which all the particle motions are reversed. Such a time-reversed system would evolve away from equilibrium. The former type of evolution can be observed, the latter cannot, but both appear to be compatible with the laws of physics. Boltzmann's well known answer was to say: 'reverse them then!' The idea behind this is that there is something special about the final high-entropy state which makes it a most improbable starting point for the time-reversed evolution, namely, intricate correlations exist between the particle velocities, which would lead to the emergence of structure from apparent equilibrium in the time-reversed case. These correlations result from the fact that the state is taken to arise from a low-entropy state in the past (Penrose 1979). The evolution of a physical system is therefore governed by the 'thermodynamic' arrow of time. Of course, we have not explained why the thermodynamic arrow should operate in this way; i.e. why low-entropy states should be suitable initial states.

Various attempts have been made to link the thermodynamic arrow with other apparent determinants of a direction for the flow of time. Consider first the arrow apparently provided by electrodynamics; the point usually made can be seen in an example. A radio programme is the result of articulate speakers causing electronic apparatus to vibrate and emit radio waves into the future. The time-reversed situation, in which radio waves impinge on a 'receiver' and cause performers to speak, is not observed. But this is merely another aspect of the thermodynamic arrow, according to which the low-entropy state prior to the broadcast causes the correlations in the radio waves at a later, high-entropy stage. One might be tempted to argue that there is an additional electrodynamic arrow provided by the fact that while we see sink-free electromagnetic radiation emitted into the future, we never see source-free radiation from the past. That is, if a telescope receives a photon we immediately try to find its source, and are apparently unhappy with the possibility that it does not have a source; whereas we have no qualms about contemplating a photon emitted to cross the Universe for ever. However, it would in fact make no physical difference to our Universe if some of the radiation we receive were simply source-free radiation created along with the matter. So this version of the electrodynamic arrow is not relevant in any important way.

Consider next the 'cosmological arrow' supposedly provided by the expansion of the Universe. Certainly a universe in stable equilibrium would have no sense of time. In addition, the expansion provides a welcome large sink for the radiation of the stars, which can therefore evolve in the standard way. But nothing would be changed immediately if the expansion were stopped; in particular, the stars would continue to shine outwards, and the local thermodynamic arrow would continue to operate. There is no reason to suppose that during the recollapse phase of a closed universe time would run backwards. The observer would simply see the stars shining into a partly blue-shifted sky. Since thermodynamics could proceed normally in a contracting universe, there is no obvious link between the thermodynamic arrow of time and that provided by universal expansion. It is conceivable that a long-range interaction with the Universe links the cosmological and thermodynamic arrows, but this is pure speculation.

An alternative view is that time irreversibility might be present in as yet unknown local physical laws. There is some evidence for this in the decay of the K^o and $\bar{K}^o$ mesons. Whilst other elementary processes give rise to equivalent sets of processes if time is reversed, on the basis of deductions from its observed decay modes the K^o-$\bar{K}^o$ system seems to violate this. The existence of a (very weak) time reversal non-invariant interaction is presumably responsible. However, the link, if any, with the thermodynamic arrow is obscure.

The behaviour of gravitating systems with regard to entropy is particularly interesting. Gravitating systems are unstable to condensations, as is clearly demonstrated in the evolution of the Universe itself. The end-point of the operation of

245

gravitation is a state of maximum condensation. Thus stars evolve into black holes; black holes grow by swallowing more material and radiation, and decay by the quantum emission of radiation. The ever-expanding universe models end in the conversion of all matter to an infinite sea of radiation expanding to zero temperature. In the recollapsing models, black holes grow and coalesce as the Universe recontracts, and the radiation rises to infinite temperature at the final singularity. In all cases all the carefully built up gravitational structure is finally dissipated in a state of maximum entropy. Note that the formation of black holes means that the future singularities on the Universe (inside the black holes) are different from those in the past (at the big bang) and this could provide another link to a cosmological arrow of time.

From this point of view the $10^9 k$ of entropy per baryon in the hot big bang represents a *low-entropy* initial state. The operation of gravity produces condensations and regions of sufficient negative entropy in which the Universe can, apparently, be understood. But the evolution of structure demands an arrow of time, and this arrow seems necessarily to require the dissolution of structure into high entropy. The Universe, it would appear, evolves to just that state in which it can know its own oblivion. Throughout all the galaxies, on countless shores of fragile green, countless intelligences discover the Universe to be merely a joke. This, it seems, is the vision featured in the patterns of all those stars, and in the patterns of the patterns, from which we fashion a Universe amidst its black amnesias.

References

Abell G O 1958 *Astrophys. J. Suppl.* **3** 211
—— 1975 in *Galaxies and the Universe* ed G Sandage (University of Chicago Press)
Allen C W 1972 *Astrophysical Quantities* 3rd edn (London: Athlone Press)
Bachall N 1977 *Ann. Rev. Astron. Astrophys.* **15** 505
Barrow J 1976 *Mon. Not. R. Astron. Soc.* **175** 359
Barrow J and Carr B 1978 *Mon. Not. R. Astron. Soc.* **182** 537
Barrow J and Matzner R 1977 *Mon. Not. R. Astron. Soc.* **181** 719
Baum W A 1972 in *IAU Symp. No. 44, External Galaxies and Quasi-Stellar Objects* ed D S Evans (Dordrecht: Reidel) p393
Binney J 1977 *Astrophys. J.* **215** 483
Birkinshaw M, Gull S F and Northover K J E 1978 *Mon. Not. R. Astron. Soc.* **185** 245
Blackman R B and Tukey J W 1959 *The Measurement of Power Spectra* (New York: Dover)
Bondi H 1947 *Mon. Not. R. Astron. Soc.* **107** 410
—— 1960 *Cosmology* (London: Cambridge University Press)
Braginsky V B and Panov V I 1972 *Sov. Phys.–JETP* **34** 463
Bruzual G A and Spinrad H 1978 *Astrophys. J.* **220** 1; **222** 1119
Burbidge G R, Crowne A H, Smith H E and Harding E 1977 *Astrophys. J. Suppl.* **33** 113
Candelas P and Raine D J 1975 *Phys. Rev.* **D12** 965
Chandrasekhar S 1960 *Radiative Transfer* (New York: Dover)
Cheng E S, Saulson P K, Wilkinson D T and Corey B E 1979 *Astrophys. J. Lett.* **232** L139
Colla G, Fanti C, Fanti R, Ficarra A, Formiggini L, Gandolfi E, Lari C, Marano B, Padrielli L and Tomasi P 1972 *Astron. Astrophys. Suppl.* **7** 1
Collins C B and Ellis G F R 1979 *Phys. Rep.* **C56** 65
Collins C B and Hawking S W 1973a *Astrophys. J.* **180** 317
—— 1973b *Mon. Not. R. Astron. Soc.* **162** 307
Collins C B and Stewart J M 1971 *Mon. Not. R. Astron. Soc.* **153** 419
Davis M 1976 in *Frontiers of Astrophysics* ed E H Avrett (Harvard University Press)
Davis M, Groth E J and Peebles P J E 1977 *Astrophys. J.* **212** L107
D'Eath P D 1976 *Ann. Phys. (NY)* **98** 237
De Veny J B, Osborn W H and Janes K 1971 *Publ. Astron. Soc. Pacific* **83** 611
Dicke R H 1964 *The Theoretical Significance of Experimental Relativity* (London: Blackie)
Dicke R H, Peebles P J E, Roll P G and Wilkinson D T 1965 *Astrophys. J.* **142** 414

Dyer C C 1976 *Mon. Not. R. Astron. Soc.* **175** 429

Ellis G F R 1971 in *General Relativity and Cosmology, Proc. Int. School of Physics 'Enrico Fermi', Course 47* ed R K Sachs (New York: Academic Press)

—— 1975 *Ann. NY Acad. Sci.* **262** 231

Fabbri R, Guidi I, Melchiorri F and Natale V 1980 *Phys. Rev. Lett.* **44** 1563

Fabian A C and Sanford R W 1971 *Nature Phys. Sci.* **231** 52; **234** 20

Fall S M 1976 *Mon. Not. R. Astron. Soc.* **176** 181

—— 1979 *Rev. Mod. Phys.* **51** 21

Gilman R C 1970 *Phys. Rev.* **D2** 1400

Gott J R, Gunn J E, Schramm D N and Tinsley B M 1974 *Astrophys. J.* **194** 543

Groth E J and Peebles P J E 1977 *Astrophys. J.* **217** 385

Grueff C and Vigotti M 1977 *Astron. Astrophys.* **54** 475

Gull S and Northover K 1976 *Nature* **263** 572

Gunn J E and Oke J B 1975 *Astrophys. J.* **195** 255

Hauser M G and Peebles P J E 1973 *Astrophys. J.* **185** 757

Hawking S W 1969 *Mon. Not. R. Astron. Soc.* **142** 129

Hawking S W and Ellis G F R 1973 *The Large Scale Structure of Space–Time* (London: Cambridge University Press)

Hoyle F 1953 *Astrophys. J.* **118** 513

Hulse R A and Taylor J H 1975 *Astrophys. J.* **195** L51

Jaakkola T, Karoji H, Ledenmat G, Moles M, Nattale L, Vigier J P and Pecker J C 1976 *Mon. Not. R. Astron. Soc.* **177** 191

Jeffreys H 1931 *Cartesian Tensors* (London: Cambridge University Press)

Jones B J T 1976 *Rev. Mod. Phys.* **48** 107

Kirschner R P and Kwan J 1974 *Astrophys. J.* **193** 27

Kristian J and Sachs R K 1966 *Astrophys. J.* **143** 379

Kristian J, Sandage A and Westphal J A 1978 *Astrophys. J.* **221** 383

Layzer D 1975 in *Galaxies and the Universe* ed A Sandage *et al* vol 9 of *Stars and Stellar Systems* (University of Chicago Press)

Lightman A P and Lee D L 1973 *Phys. Rev.* **D8** 364

Longair M S and Einasto J (eds) 1978 *The Large-scale Structure of the Universe, IAU Symp. No. 79* (Dordrecht: Reidel)

Lyubimov V A, Novikov E G, Nozik V Z, Tretyakov E F and Kosik V S 1980 unpublished

MacCallum M A H 1973 in *Cargèse Lectures in Physics* vol 6 ed E Schatzman (London: Gordon and Breach)

—— 1979 in *General Relativity, an Einstein Centenary Survey* ed S W Hawking and W Israel (London: Cambridge University Press)

Mach E 1883 *Die Mechanik in Ihrer Entwicklung Historische-Kritische Dargestellt* transl. by T J McCormack (1960) as *The Science of Mechanics* (Open Court)

Markarian B Ye and Lipovetsky V A 1974 *Astrofizika* **10** 307

Marshall F E, Boldt E A, Holt S S, Miller R B, Mushotzky R F, Rose L A, Rothschild R E and Serlemitsos P J 1980 *Astrophys. J.* **235** 4

Misner C W 1969 *Phys. Rev. Lett.* **22** 1071

Moffett A T 1975 in *Galaxies and the Universe* ed A Sandage *et al* vol 9 of *Stars and Stellar Systems* (University of Chicago Press)

Morgan W W 1958 *Publ. Astron. Soc. Pacific* **70** 364

—— 1959 *Publ. Astron. Soc. Pacific* **71** 394

Muller R A 1978 *Sci. Am.* **238** 64

North J D 1965 *Measure of the Universe* (Oxford University Press)

Omnès R 1969 *Phys. Rev. Lett.* **23** 38

—— 1971 *Astron. Astrophys.* **11** 450

—— 1972 *Phys. Rep.* **3C** 1

Page D 1975 in *Galaxies and the Universe* ed A Sandage *et al* vol 9 of *Stars and Stellar Systems* (University of Chicago Press)

Pariskij Yu N 1973 *Sov. Astron. Astrophys. J.* **17** 219

Partridge R B 1980 *Astrophys. J.* **235** 681

Peebles P J E 1971 *Physical Cosmology* (Princeton University Press)

—— 1975 *Astrophys. J.* **196** 647

Peimbert M 1973 *IAU Symp. No. 58* ed J R Shakeshaft (Dordrecht: Reidel)

Penrose R 1979 in *General Relativity, An Einstein Centenary Survey* ed S W Hawking and W Israel (London: Cambridge University Press)

Penzias A A and Wilson R W 1965 *Astrophys. J.* **142** 419

Perrenod S D and Lada C J 1979 *Astrophys. J.* **234** L173

Pooley G G and Ryle M 1968 *Mon. Not. R. Astron. Soc.* **139** 515

Pratt N M 1977 *Vistas in Astronomy* **21** 1

Raine D J 1975 *Mon. Not. R. Astron. Soc.* **171** 507

Raine D J and Heller M 1981 *The Science of Space–Time* (London: Pachart)

Rees M J 1972 *Phys. Rev. Lett.* **28** 1669

—— 1978 *Nature* **275** 35; 343

Rees M J, Ruffini R and Wheeler J A 1974 *Black Holes, Gravitational Waves and Cosmology* (New York: Gordon and Breach)

Rees M J and Ostriker J P 1977 *Mon. Not. R. Astron. Soc.* **179** 541

Rees M J and Sciama D W 1968 *Nature* **217** 511

Reines F W, Scobel H W and Pasierb E 1980 unpublished

Robertson J G 1973 *Australian J. Phys.* **26** 403

Robson E I, Vickers D G, Huizinga J S, Beckman J E and Clegg P E 1974 *Nature* **251** 591

Rudnicki K, Dworak T Z, Flin P, Baranowski B and Sendrakowski A 1973 *Acta Cosmologica* **1** 7

Sachs R K and Wolfe A M 1967 *Astrophys. J.* **147** 73

Sandage A 1961 *The Hubble Atlas of Galaxies* (Washington: Carnegie Institute)

—— 1972 *Astrophys. J.* **178** 1

Sandage A and Tammann G A 1975 *Astrophys. J.* **197** 265

Schmidt M 1968 *Astrophys. J.* **162** 371

Schwarz D 1970 *Astrophys. J.* **162** 439

Sciama D W 1959 *Unity of the Universe* (London: Faber and Faber)

Sciama D W, Waylen P G and Gilman R C 1969 *Phys. Rev.* **D187** 1762

Seldner M and Peebles P J E 1977 *Astrophys. J.* **215** 703

Shane C D and Wirtanen C A 1967 *Publ. Lick Observatory* **22** 1

Shapley H and Ames A 1932 *Ann. Harvard College Observatory* **88** 43

Silk J 1977 *Astrophys. J.* **211** 638

Smith M G 1978 *Vistas in Astronomy* **22** 321

Smoot G F, Gorenstein M V and Muller R A 1977 *Phys. Rev. Lett.* **39** 898

Smoot G F and Lubin P H 1979 *Astrophys. J. Lett.* **234** L83

Steigman G 1973 in *Cargèse Lectures in Physics* vol 6 ed E Schatzman (New York: Gordon and Breach)

Stewart J M 1968 *Astrophys. Lett.* **2** 133

Taylor J H, Fowler L A and McCulloch P M 1979 *Nature* **277** 437

Thorne K S, Lee D L and Lightman A P 1973 *Phys. Rev.* **D7** 3563

Turner M S and Schramm D N 1979 *Physics Today* **32** 42

Van den Bergh 1959 *Publ. David Dunlap Observatory* vol 2 No. 5

—— 1960 *Publ. David Dunlap Observatory* vol 2 No. 6

de Vaucouleurs G and Bollinger G 1979 *Astrophys. J.* **233** 433

de Vaucouleurs G and de Vaucouleurs A 1964 *Reference Catalogue of Bright Galaxies* (University of Texas Press)

Wagoner R V 1974 in *Confrontation of Cosmological Theories with Observational Data, IAU Symp. No.* 63 ed M S Longair (Dordrecht: Reidel)

Warwick R S, Pye J P and Fabian A C 1980 *Mon. Not. R. Astron. Soc.* **190** 243

Webster A 1976 *Mon. Not. R. Astron. Soc.* **175** 61; 71

Weedman D W 1977 *Ann. Rev. Astron. Astrophys.* **15** 69

Weinberg S 1972 *Gravitation and Cosmology* (New York: Wiley)

—— 1977 *The First Three Minutes* (London: Andre Deutsch)

Will C M 1974 in *Experimental Gravitation, Proc. Course 56 Int. School of Physics 'Enrico Fermi'* ed B Bertotti (New York: Academic Press)

—— 1979 in *General Relativity, An Einstein Centenary Survey* ed S W Hawking and W Israel (London: Cambridge University Press)

Woody D P, Mather J C, Nishioka N S and Richards P L 1975 *Phys. Rev. Lett.* **34** 1036

Woody D P and Richards P L 1979 *Phys. Rev. Lett.* **42** 925

Zeldovich Ya B 1972 *Mon. Not. R. Astron. Soc.* **160** 1P

Zeldovich Ya B and Sunyaev R A 1969 *Astrophys. Space Sci.* **4** 301

Zwicky F, Herzog E, Wild P, Karpowicz M and Kowal C T 1961–1968 *Catalogue of Galaxies and Clusters of Galaxies* (6 vols) (California Institute of Technology)

Index